SpringerBriefs in Earth Sciences

For further volumes:
http://www.springer.com/series/8897

V. F. Pisarenko · M. V. Rodkin

Statistical Analysis of Natural Disasters and Related Losses

Springer

V. F. Pisarenko
M. V. Rodkin
International Institute of Earthquake
Prediction Theory and Mathematical Geophysics
Russian Academy of Sciences
Moscow
Russia

ISSN 2191-5369 ISSN 2191-5377 (electronic)
ISBN 978-3-319-01453-1 ISBN 978-3-319-01454-8 (eBook)
DOI 10.1007/978-3-319-01454-8
Springer Cham Heidelberg New York Dordrecht London

Library of Congress Control Number: 2013944884

Printed on acid-free paper

Springer is part of Springer Science+Business Media (www.springer.com)

Preface

The past decades demonstrate a spectacular increase of public interest in the problems of safety and reduction of losses from natural and manmade disasters. The study of disaster statistics and disaster occurrence is a complicated interdisciplinary field involving an intimate interplay of new theoretical results from several branches of mathematics, physics, and computer science. Recent progress in systems of remote sensing, in financial and social measurements, and in data collecting gives us a possibility to compile data sets suitable for statistical analysis of disaster damages. The accumulation of factual material relating to various kinds of natural disasters and the use of advanced recording techniques have expanded possibilities for the analysis of empirical distributions of disaster characteristics. This book summarizes recent achievements in the field of statistical analysis of disaster damages. These approaches largely rely on fundamental results of the theory of extreme values (Embrechts et al. 1997; Gumbel 1958; Pisarenko and Rodkin 2010). One should mark the appearance of theoretical and practical tools for effective studies of natural disasters with the associated practical measures taken to reduce the losses. The combination of all the above-mentioned factors results in a considerable progress in natural disaster research. The main focus is on the occurrence of disasters that can be described by distributions with heavy tails. A short overview of properties of heavy-tailed distributions is offered. The relation between the maximum event and the total sum is studied; in the case of heavy-tailed distributions this relation is unusual: the single maximum event can dominate the total sum. This book contains several recent results in the statistical analysis of rare large events. We analyze the size distribution of arbitrary nature in the uppermost range of extremely rare events using a recently developed method (Pisarenko and Rodkin 2010, 2013). One of the most important results of this study is the conclusion about instability of the "maximum possible size" parameter, this parameter is frequently used in seismic risk assessment and in other similar problems. We suggest an alternative robust way to parameterize the tail of the size distribution by means of a robust and stable characteristic—the quantiles $Q_q(\tau)$ of maximum size (e.g., earthquake energy, ground acceleration caused by earthquake, victims and economic losses from natural catastrophes, etc.) that will occur in a prescribed time interval τ. We illustrate our theoretical conclusions by applying the described technique to different natural disasters. The comparative study of losses from earthquakes, floods, tornsdoes and hurricanes is presented. The losses of different types are analyzed: fatalities, number of affected people, overall economic losses. We also emphasize that the methods used here to parameterize the distribution's tail are quite general and are applicable besides the assessment of natural disaster hazards to all the cases where the contribution of rare large events is to be estimated.

The book presents an original approach in the field of disaster statistics recently developed in (Embrechts et al. 1997; Gumbel 1958; Pisarenko and Rodkin 2010). We analyze empirical data related to several types of natural disasters: earthquakes, floods, tornsdoes and hurricanes. The mode of occurrence and statistics of losses from such disasters are considered in detail.

This book aims primarily at specialists in the field of seismology and seismic risk, and could also be useful for specialists in other kinds of natural and manmade disasters. The main statistical results are derived with a mathematical rigor and are presented here in a form that also makes them accessible to readers with no special mathematical background. We hope that this monograph will also be useful for employees of regional and national administrations as well as for a broad class of readers interested in the problems of natural disasters and their impact on the society.

This book has the following structure. Chapter 1, the *Heavy-Tailed Distributions and Their Properties* provides a general overview of heavy-tailed distributions and their properties. In Chap. 2, *The Stable Approach to the Risk Assessment: Estimation of Quantiles of Maximum Event* we describe the specific properties of heavy-tailed distributions and the stable approach to the risk assessment based on the estimation of quantiles of maximum event. We analyze among others the problem of non-stationarity of natural processes (Sect. 2.2) which is important in numerous applications. Several statistical tools are suggested for dealing with this problem. In Sect. 2.4, we adopt our method to the format of aggregated annual data. Chapter 3 is devoted to the statistical analysis of real catalogs of natural disasters: earthquakes, floods, tornadoes, and hurricanes. The quantiles $Q_q(\tau)$ are estimated for each of these catalogs. Finally, in Chap. 4 we discuss the obtained results and give a comparative overview of the analyzed catalogs.

Moscow, June 2013

V. F. Pisarenko
M. V. Rodkin

References

Embrechts P, Klueppelberg C, Mikosch T (1997) Modelling extremal events. Springer, Berlin
Gumbel EJ (1958) Statistics of extremes. Columbia University Press, New York
Pisarenko VF, Rodkin MV (2010) Heavy-tailed distributions in disaster analysis. Springer, New York
Pisarenko VF, Rodkin MV (2013) The new quantile approach: application to the seismic risk assessment. In: Rascobic B, Mrdja S (eds) Natural disasters: prevention, risk factors and management. Nova Publishers, New York, pp 141–174

Acknowledgments

The authors are thankful to Prof. F. Aptikaev for very useful discussions. They are also grateful to Dr. D. V. Pisarenko who has read the manuscript and gave very useful remarks.

Contents

Chapter 1
Heavy-Tailed Distributions and Their Properties

Abstract We define the heavy-tailed distribution as distribution with infinite mathematical expectation. For such distributions the standard statistical tools—sample mean and sample standard deviation—exhibit a high instability. Some examples illustrating this conclusion are presented. We discuss relations between the maximum event $M_{max}^{(n)} = max(x_1, \ldots, x_n)$ and the total sum $S_n = x_1 + \cdots + x_n$ for heavy-tailed distributions. The asymptotical proportionality of $M_{\max}^{(n)}$ and S_n is derived for the heavy-tailed Pareto distributions (Eq. (1.18)).

Keywords Heavy tailed distributions · Pareto distribution · Maximum event—total sum relations

The main focus of our study is made on damages from disasters whose distribution can be described by power-like laws with a heavy tail. The disasters of this type typically occur in a very broad range of scales, the rare greatest events being capable of causing losses comparable with the total loss due to all the other (smaller) events. The disasters of such a type are the most unexpected ones and they frequently entail huge losses. One of the most important results of our study is the conclusion about the instability of the "maximum possible earthquake" parameter $M_{\max}$ which is frequently used in the seismic risk assessment. Instead of focusing on the unstable parameter $M_{\max}$ we suggest a stable and convenient characteristic $M_{\max}(\tau)$ defined as the maximum size of a given phenomenon that can be recorded over a future time interval τ. The random value $M_{\max}(\tau)$ can be described by its distribution function, or equivalently by its quantiles $Q_q(\tau)$, which are stable, robust characteristics in contrast to $M_{\max}$. Besides, if $\tau \to \infty$, then $M_{\max}(\tau) \to M_{\max}$ with probability one. The method of calculation of $Q_q(\tau)$ is exposed below. In particular, we can estimate $Q_q(\tau)$ for, say, $q = 90$, 95 and 99 %, as well as for the median ($q = 50$ %) for any desirable time interval τ. Our method provides an alternative and robust way to parameterize the rightmost tail of the frequency-size relation. The final goal of our data processing consists in obtaining a family of quantiles $Q_q(\tau)$ that corresponds to damage (fatalities in a disaster, economic damage, insurance losses etc.) that will not be exceeded in future τ years

V. F. Pisarenko and M. V. Rodkin, *Statistical Analysis of Natural Disasters and Related Losses*, SpringerBriefs in Earth Sciences,
DOI: 10.1007/978-3-319-01454-8_1,

with a prescribed probability q. This characteristic seems to be particularly useful, for instance in the insurance business.

It is well known that distributions of some parameters of natural processes (earthquake energy, economic damage and casualties from natural disasters and others) are often modeled by power-like laws, such as the Pareto distribution. The Pareto distribution function and probability density are given as follows:

$$F(x) = 1 - (h/x)^{\beta};\ f(x) = \frac{\beta \cdot h^{\beta}}{x^{1+\beta}};\quad x \geq h. \tag{1.1}$$

If the exponent β of such a distribution is less than unity, $\beta \leq 1$, then the mathematical expectation of the corresponding random variable is infinite. In this case, the standard statistical tools, such as sample mean and sample standard deviation, exhibit a high instability. Such distributions are often called distributions with *heavy tails*. It should be noted here that there is no commonly accepted definition of a heavy tailed distribution. Here we apply this term to distributions $F(x)$ whose mathematical expectation is infinite:

$$Ex = \int xdF(x) = \infty. \tag{1.2}$$

Accordingly, the opposite term *light tail* will designate distributions with finite expectation.

However there is a wide class of distributions where mathematical expectation is finite but higher statistical moments are infinite. Such distributions might be named "heavy-tailed" as well.

One may note that there are alternative definitions of the terms "heavy tail" and "light tail", see Reiss and Thomas (1997), Embrechts et al. (1997).

The Pareto law is a classical example of a heavy tailed distribution provided that $\beta <= 1$. In that case its expectation is infinite, so that the Law of Large Numbers is inapplicable, and the sample mean and the sample standard deviation are unstable. In contrast with a more common case of distributions with finite expectation, the increase of sample size does not improve the accuracy of the sample mean. The large statistical scatter of values of the sample mean (see e.g. Osipov 2002) makes this widely used statistical characteristic inadequate for applications to data sets with heavy tailed distributions. Moreover, its use in such situations may lead to essential errors and incorrect conclusions. Hereafter we are giving an illustration to this affirmation.

The total number of fatalities from typhoons, hurricanes and floods in the world during the period from 1947 till 1960 is about 900,000 (the flood causing the highest fatality of 1,300,000 occurred in China in 1931). The mean annual fatality (arithmetic mean) for the period from 1947 till 1960 was $\hat{X}_{ann} = 64{,}300$. The mean annual fatality for the period from 1962 till 1992 is $\hat{X}_{ann} = 36{,}000$. These sample means differ significantly and provide no indication about the probability of such a super-catastrophic event as the 1931 flood in China.

Let us consider another example: the total number of fatalities from earthquakes in the world during the period from 1947 till 1970 was about 151,000 (the earthquake causing the highest death toll of 240,000 occurred in China in 1976). The mean annual fatality rate (arithmetic mean) for the period from 1947 till 1970 is $\hat{X}_{ann} = 6{,}300$. The mean annual fatality rate for the period from 1962 till 1992 is $\hat{X}_{ann} = 18{,}600$. Here again, both sample means give quite an uncertain characteristic of annual fatality, which is a typical indication of a data set with the heavy-tailed distribution.

Figure 1.1 shows the non-normalized complementary sample distribution function for: *a*—annual casualties; *b*—number of victims in a single event; *c*—annual economic losses. These data are taken from the Web site of the US Geological Survey (http://www.neic.cr.usgs.gov/neis/eqlists). The Pareto law fits to the middle range and to the extreme events range are shown by dotted lines, with $\beta = 0.77 \pm 0.11$ (a); $\beta = 0.73 \pm 0.11$ (b); $\beta = 0.65 \pm 0.16$ (c). We can see that all the three distributions on Fig. 1.1 can be qualified as heavy-tailed.

Below we present some new approaches to the reliable statistical estimation in the situations where the standard statistical tools fail.

In the statistical analysis of data with heavy-tailed distributions an important role belongs to the maximum observed event $M_{\max}^{(n)} = max(x_1, \ldots, x_n)$. It can be shown that for a data set with a heavy tailed distribution (according to definition (1.2)) the maximum event $M_{\max}^{(n)}$ is commensurable with the total sum $S_n = x_1 + \cdots + x_n$, i.e. $M_{\max}^{(n)}$ and S_n are of the same order. For example, in the case of the Pareto distribution with exponent $\beta < 1$ one has:

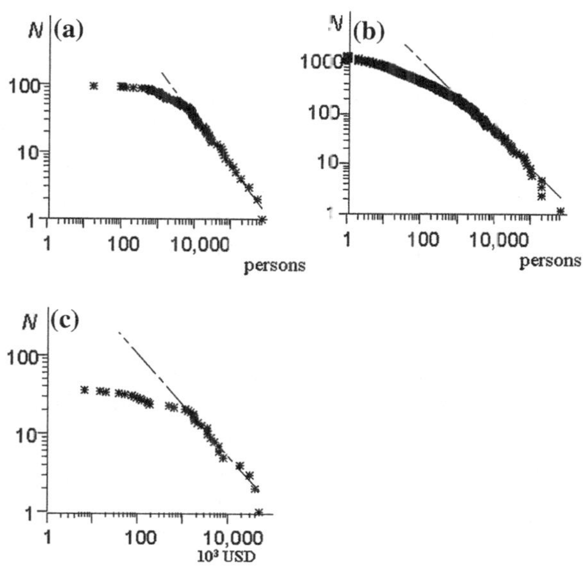

Fig. 1.1 Non-normalized complementary sample of world-wide losses from earthquakes distribution function for: **a** annual casualties; **b** casualties for individual event; **c** annual economic losses. The power-law functions fitting tails are shown by dotted lines, with $\beta = 0.77 \pm 0.11$ (**a**); $\beta = 0.73 \pm 0.11$ (**b**); $\beta = 0.65 \pm 0.16$ (**c**)

$$E\frac{S_n}{M_{\max}^{(n)}} \to \frac{1}{1-\beta}, \quad n \to \infty \tag{1.3}$$

Conversely, for a data set with a light tailed distribution (in which case the expectation is finite $E|x| < \infty$), the contribution of $M_{\max}^{(n)}$ in the total sum S_n decreases with the sample size n:

$$\frac{M_{\max}^{(n)}}{S_n} \to 0, \quad n \to \infty \tag{1.4}$$

If a sample is characterized by the Pareto distribution with index β, then $M_{\max}^{(n)}$ grows with the size of the sample as $n^{1/\beta}$. (We would like to emphasize that this tendency of a non-linear growth of $M_{\max}^{(n)}$ with the sample size n or, equivalently, with the observation time span τ can be incorrectly interpreted as an evidence of a non-stationarity in time (Pisarenko and Rodkin 2010). In many cases, the wide-spread belief that the rate of losses from natural disasters has a clear tendency of increasing with time is precisely connected with the misinterpretation of this apparent non-stationarity effect).

The treatment of heavy-tailed data is often facilitated by using logarithms of original values. Switching to logarithms (which can be done only when the original numerical values are positive) ensures almost always that all the statistical moments exist, and hence the Law of Large Numbers and the Central Limit Theorem are applicable to the sums of logarithms. We are advocating the use of the logarithmic transform of the original data values as one of the main tools to prevent complications in the processing of heavy-tailed data sets. As we shall see later, the use of the logarithmic transform turned to be efficient in all the considered cases except one (the data set on the fatality of floods in USA from 1995 till 2011).

It should be remarked that if X has the Pareto distribution, then $\log(X)$ has exactly exponential distribution. We shall use this fact below.

Another statistical tool helping to overcome difficulties connected with processing heavy tailed data is *the order statistics*: sample median, sample quantiles, interquartile range etc. The order statistics can be used also for construction of confidence intervals. The main statistical tool suggested in this short-monograph for description of distribution tail—the family of quantiles $Q_q(\tau)$—is in fact a continuous analog of the sample ordered statistics. This explains its robustness and stability.

In our approach we model the sequence of disaster occurrences by well-known Poisson point process, see Embrechts et al. (1997). The disaster effects can be different: fatalities and economic losses due to natural catastrophes, earthquake seismic moments and ground acceleration at a fixed point etc. These effects are "marks" assigned to occurrence times of the point Poisson process. Thus, the whole construction is called the marked point process. In our applications we shall model our catalog usually by a stationary Poisson process with intensity λ events per year (but we devote the special section *Non-stationarity of natural processes, "operational time" method* to the case of non-stationary Poisson processes). The random number of events in stationary process for T years is a random Poisson variable with the mean λT. Let us denote the maximum event for this time period as M_T and the total sum as $\sum_T$. We are now going to derive some relations between M_T and $\sum_T$. It is well known (see, e.g., Gumbel 1958) that for distributions with a light, exponential tail M_T increases as the logarithm of T:

$$M_T \cong c \cdot \log(\lambda T), \tag{1.5}$$

where c is some constant, whereas $\sum_T$ increases as T, in accordance with the CLT:

$$\sum\nolimits_T \cong \lambda T \cdot b + \zeta \cdot \sigma \cdot (\lambda T)^{1/2}, \tag{1.6}$$

where ζ is some standard normal rv, b is the expectation of a single event (here we assume $b > 0$), and σ is the standard deviation (std) of a single event. We consider the ratio $R(T)$ of the total sum to maximum event:

$$R(T) = \sum\nolimits_T / M_T.$$

It follows from (1.5), (1.6) that

$$R(T) \cong (b/c) \cdot (T/\log(\lambda T)). \tag{1.7}$$

Hence it follows that the $R(T)$ ratio increases with T linearly for the cases of light tail if we disregard the slowly varying $\log(\lambda T)$. A quite different behavior of $R(T)$ appears for heavy-tailed distributions. For such distributions $R(T)$ grows much slower and can even have a finite expectation (cf. with (1.3)). In other words, in this case the total and the maximum event become comparable, i.e., the total sum is determined in a large extent by the single maximum event.

In order to illustrate this assertion, we are going to derive lower and upper bounds for $R(T)$ corresponding to the Pareto distribution. Taking logarithms and their expectation we get

$$E \log \sum\nolimits_T = E \log R(T) + E \log M_T \tag{1.8}$$

Using Jensen's inequality for concave functions, we get:

$$E \log R(T) \leq \log ER(T). \tag{1.9}$$

As shown in Pisarenko (1998),

$$ER(T) = \begin{cases} [1 - (\lambda \mathrm{T})^{1-1/\beta} \cdot \gamma(1/\beta; \lambda \mathrm{T})]/(1-\beta); & \beta \neq 1; \quad (1.10) \\ \log(\lambda T) - \exp(-\lambda T) \cdot (\log \lambda T - 1); & \beta = 1, \quad (1.11) \end{cases}$$

Table 1.1 Expectation $E\ R(T)$ as function of λT and β

β	λT (or n)						
	10	50	100	300	500	1000	∞
3	5.77	17.93	28.8	60.7	85.7	136.6	∞
2	4.6	11.52	16.7	29.7	38.6	55	∞
1.3	3.48	6.55	8.26	11.6	13.5	16.42	∞
1.1	3.1	5.16	6.15	7.85	8.7	9.91	∞
1.0	2.88	4.49	5.18	6.28	6.78	7.48	∞
0.9	2.59	3.81	4.26	4.92	5.2	5.56	10
0.8	2.45	3.29	3.57	3.91	4.04	4.19	4.3
0.6	2.02	2.33	2.4	2.45	2.46	2.48	2.5

where $\gamma(\alpha; t)$ is the incomplete gamma function. The function $ER(T)$ is tabulated in Table 1.1.

If $\lambda T \to \infty$, then in (1.10) the incomplete gamma function $\gamma(1/\beta; \lambda T)$ tends to the standard gamma function $\Gamma(1/\beta)$. It is possible to derive an explicit expression for $E \log(M_T)$ (Pisarenko 1998):

$$E \log M_T = [\log(\lambda T) + C - Ei(-\lambda T)/\beta] \tag{1.12}$$

where C is the Euler constant ($C = 0.577\ldots$), and $Ei(-\lambda T)$ is the integral exponential function

$$Ei(x) = \int_{-\infty}^{x} \frac{\exp(z)}{z} dz; \quad x<0.$$

Combining Eqs. (1.8)–(1.12), we derive an upper bound on $E \log \Sigma_T$:

$$E \log \sum\nolimits_T \le \log \frac{1 - (\lambda T)^{1-1/\beta}\gamma(1/\beta; \lambda T)}{1-\beta} + [\log(\lambda T) + c - Ei(-\lambda T)]/\beta. \tag{1.13}$$

If $\lambda T \gg 1$, then (1.13) can be simplified. Keeping only terms growing with (λT), we get:

$$E \log \sum\nolimits_T \le max\left(1; \frac{1}{\beta}\right) \cdot \log(\lambda T) \tag{1.14}$$

Now we derive a lower bound for $E \log\Sigma_T$. *If* $\beta > 1$ then by virtue of the Law of Large Numbers $\frac{1}{n}\Sigma_T \to 1/\beta$ and we get:

$$E \log \sum\nolimits_T = E \log\left(n \frac{1}{n} \sum\nolimits_T\right) = E \log(n) + 1/\beta \cong \log(\lambda T) + 1/\beta \tag{1.15}$$

If $\beta < 1$, then we just drop the first term in rhs of (1.8) and get:

$$E \log \sum\nolimits_T > E \log M_T = [\log(\lambda T) + C - Ei(-\lambda T)/\beta] \qquad (1.16)$$

We can put down (1.15) and (1.16) in one relation that holds true for any β and disregards terms of lower order:

$$E \log \sum\nolimits_T \geq \max\left(1; \frac{1}{\beta}\right) \cdot \log(\lambda T) \qquad (1.17)$$

We get from (1.14) and (1.17) that asymptotically (as $\lambda T \to \infty$) up to terms of lower order for any β:

$$E \log \sum\nolimits_T = \max\left(1; \frac{1}{\beta}\right) \cdot \log(\lambda T). \qquad (1.18)$$

Thus, one can say that for $\beta < 1$ random quantities $\log\Sigma_T$ and $\log M_T$ are comparable in value and both grow as $\log(\lambda T)^{1/\beta}$. This fact might be interpreted as a nonlinear growth of Σ_T and M_T at the same rate as $(\lambda T)^{1/\beta}$ does, since their logarithms are asymptotically proportional. We recall that both random quantities have infinite expectations if $\beta < 1$. If $\beta > 1$ then Σ_T increases linearly with T (cf. with (1.6)), whereas M_T increases more slowly, as $T^{1/\beta}$. The relation (1.18) holds true for any probability density that decreases asymptotically as a power.

References

Embrechts P, Klueppelberg C, Mikosch T (1997) Modelling extremal events. Springer, Berlin

Gumbel EJ (1958) Statistics of extremes. Columbia University Press, New York

Osipov VI (2002) Natural hazards control. Vestnik RAN 8:678–686 (in Russian)

Pisarenko VF (1998) Non-linear growth of cumulative flood losses with time. Hydrol Process 12:461–470

Pisarenko VF, Rodkin MV (2010) Heavy-tailed distributions in disaster analysis. Springer, Dordrecht

Reiss RD, Thomas M (1997) Statistical analysis of extreme values. Birkhauser, Basel

Chapter 2
The Stable Approach to the Risk Assessment: Estimation of Quantiles of Maximum Event

Abstract The "peak-over-threshold" method is suggested for determination of the limit distribution of the maximum event in a future time period τ, Generalized Extreme Value Distribution (GEV). This approach is based on the Extreme Value Theory, EVT. The method of Maximum Likelihood Estimation (MLE) of unknown parameters is exposed. Several statistical models of the non-stationarity of point process are studied. A modification of the suggested method for the aggregated annual data is given.

Keywords Extreme value theory · Generalized Pareto distribution, GPD · "Peak-over-threshold" method · Intensity of point process · Non-stationarity of point process · Annual data

2.1 The Method

The problem of statistical characterization of extreme, rare events is reduced to estimation of quantiles of high level (close to unity) with the estimates based on a finite sample. We recall that the quantile Q_q of level q, $0 < q < 1$ of a continuous, monotone distribution function $F(x)$ is defined as the root of the equation

$$F(x) = q.$$

Thus, the quantile q is inverse function with respect to the distribution function $F(x)$.

The problem of statistical estimation of quantiles of high level is extremely important for practice. Many applied problems boil down, in fact, to the estimation of such quantiles. If $x_1 < x_2 < \ldots < x_n$ is an ordered sample of iid (independent identically distributed) random variables with a continuous distribution function (DF) then the quantile Q_q of a fixed level q can be estimated by the kth term of the

V. F. Pisarenko and M. V. Rodkin, *Statistical Analysis of Natural Disasters and Related Losses*, SpringerBriefs in Earth Sciences,
DOI: 10.1007/978-3-319-01454-8_2, © The Author(s) 2014

ordered sample x_k (k = entire part of $q \bullet n$). This estimate is known to be consistent for fixed q as $n \to \infty$. The limit distribution of normalized sample quantile

$$n^{1/2}(x_k - Q_q)/\sigma$$

is the standard Gauss distribution; here $\sigma^2 = F(Q_q) \cdot (1 - F(Q_q))/f^2(Q_q)$; $f(x) = F'(x)$—probability density. We stress that the sample quantile tends to its theoretical analog for any distribution with continuous density whatever heavy tail is. In particular, the sample median is consistent estimate of the middle point of a symmetrical distribution (theoretical median) for any tail, whereas the sample mean (arithmetic mean of the sample values) is consistent only for distributions with a light tail. It is true that the efficiency of the sample mean is sometimes higher (e.g. for the Gauss distribution), but not much. For example, the limit standard deviation of sample mean of the standard Gauss distribution is $1/\sqrt{n}$, whereas sample median has limit standard deviation $1.25/\sqrt{n}$. The gain is not big, but the median is guaranteed against possible presence of a heavy tail component in the sample. However, if we try to estimate quantiles of higher levels, $q > 1 - 1/n$, the consistency of sample quantiles disappear. But just such levels are of the most practical interest. For example, suppose that a sample size equal to $n = 500$, so $1 - 1/n = 0.998$, whereas we need the quantile of level $q = 0.9999$ The estimation of quantiles that are "out of sample range", i.e., for $q > 1 - 1/n$, can be effectuated only under some extra assumptions about the distribution in question. There is no magical technique that would yield reliable results for free. Rephrasing a financial truth one can say:

There is no free lunch, when it comes to high quantile estimation!

We shall use for this purpose the Limit Theorem of the extreme value theory (EVT) assuming its validity (see, e.g. Embrechts et al. 1997, *Theorem 3.4.13*). The conditions guaranteeing the validity of this Limit Theorem include the regularity of the original distributions of event sizes in extreme range and boil down to the existence of a *non-degenerate* limit distribution of $\mu_n = \max(x_1, x_2, \ldots, x_n)$ after a proper centering and normalization.

If the Limit Theorem of EVT is valid, then observations exceeding a threshold h, tends (as both h and sample size n tend to infinity) to the Generalized Pareto Distribution (GPD). This approach is called sometimes the "peak over threshold" method. The GDP depends on two unknown parameters (ξ, s):

$$\begin{aligned} GPD_h(x|\xi, s) &= 1 - [1 + (\xi/s) \times (x - h)]^{-1/\xi}, \quad \xi \neq 0; \\ GPD_h(x|0, s) &= 1 - \exp(-(x - h)/s), \quad \xi = 0; \end{aligned} \tag{2.1}$$

here, ξ is the form parameter ($-\infty < \xi < \infty$), s is the scale parameter ($s > 0$). The domain of definition depends on parameter values:

$$\text{if } \xi \geq 0, \text{then } x \geq h;$$

$$\text{if } \xi < 0, \text{ then } h \leq x \leq h - s/\xi.$$

We see that for negative ξ the domain of definition of GPD is limited within a finite interval. Because of evident finiteness of any possible physical event (e.g., energy of earthquake) and of the loss values (e.g., number of fatalities) this case is mostly expected in an analysis of empirical data on parameters and losses from natural hazards, but nevertheless, sometimes the unbounded distributions can model empirical data better.

Suppose, the sample $(x_1, x_2, \ldots, x_n)$ is result of observations (peaks over a threshold h) occurred at moments $t_1, \ldots, t_n$ that represent stationary Poisson process with intensity λ. The sample is observed on time interval $[-T;\ 0]$, thus the intensity can be estimated as $\lambda = n/T$. We assume that the threshold h is high enough, so that the conditions of the Limit Theorem of EVT are fulfilled, and, consequently, observations $(x_1, x_2, \ldots, x_n)$ have GPD distribution (2.18) with some parameters (ξ, s).

We put down the formulae of the Maximum Likelihood Estimates of GPD-parameters. The GPD-density has form:

$$f(x) = \frac{1}{s}\left(1 + \frac{\xi}{s}(x - h)\right)^{-1/\xi - 1}.$$

Thus, the log-likelihood function equals:

$$L = -n \cdot \log(s) - \left(\frac{1}{\xi} + 1\right)\sum_{k=1}^{n} \log\left(1 + \frac{\xi}{s}(x_k - h)\right).$$

Now one can find numerically the ML-estimates $\hat{\xi}$, $\hat{s}$ providing maximum to L. These estimates are proved to be consistent at least for $\xi > -1/2$ and in that case the limit distribution of normalized quantities

$$\sqrt{n}\left(\hat{\xi} - \xi\right) / |1 + \xi|; \quad \sqrt{n}(\hat{s}/s - 1)\sqrt{2|1 + \xi|}; \quad \xi > -0.5$$

is the standard Gauss distribution. These relations give possibility to construct confidence intervals for parameters ξ, s.

It can be proved (Embrechts et al. 1997, Theorem 3.4.13) that if times of occurrence form a stationary Poisson process and individual sizes are GPD-distributed then the distribution of maximum $M_\tau = \max(x(t_1), \ldots, x(t_m))$ observed on interval $[0; \tau]$, $0 \leq t_1, \ldots, t_m \leq \tau$, has DF

$$\Phi_\tau(x) = \exp\left(-\lambda\tau[1 + (\xi/s) \cdot (x - h)]^{-1/\xi}\right),$$

apart from terms of the order $\exp(-\lambda\tau)$ which we assume to be negligible.

Our statistical problem consists in estimating quantiles $Q_q(\tau)$ of maximum M_τ in a future time interval τ that we propose as stable robust characteristics of the tail distribution. The quantiles $Q_q(\tau)$ are the roots of the following equation:

$$\Phi_\tau(x) = \exp\left(-\lambda\tau[1 + (\xi/s) \cdot (x - h)]^{-1/\xi}\right) = q. \tag{2.2}$$

Inverting (2.2) as a function of x depending on parameters q, τ we get:

$$Q_q(\tau) = h + (s/\xi) \cdot \left[a \cdot (\lambda\tau)^{\xi} - 1\right],$$

where $a = [\log(1/q)]^{-\xi}$

The GPD-distribution includes all types of tails: power-like ($\xi > 0$), exponential ($\xi = 0$), finite boundary ($\xi < 0$). If $\xi < 0$, then the rightmost boundary of GPD-distribution designed as $M_{\max}$ equals to:

$$M_{\max} = h - s/\xi \tag{2.3}$$

On Fig. 2.1 we show a set of GPD-tails for negative form parameters $\xi = -0.8$; -0.35; -0.1; -0.01; corresponding scale parameters s = 8.0; 3.5; 1.0; 0.1 and threshold h = 0; in all cases $M_{\max} = 10$. We see that the more absolute value $|\xi|$ is the more the tail curvature becomes and the steeper the extreme part of the tail decreases. The last curve ($\xi = -0.01$) practically coincides with the tail graph of the exponential distribution.

On Fig. 2.2 we show two GPD-densities, corresponding to negative form parameters $\xi = -0.05$ and $\xi = -0.20$. It is seen from Eq. (2.3) that the more the ratio $-s/\xi$ is, the further to right side $M_{\max}$ is shifted. The density looks like a duck beak. This explains instability of the parameter $M_{\max}$: small variation of ξ-estimate can lead to large excursion of $M_{\max}$.

The main difficulty in parameter estimation of GPD consists in the choice of a proper threshold h. How to cut off the utmost part of the tail for further analysis and statistical inference about asymptotical tail behavior? We use for this purpose the Kolmogorov test with corrections for estimated parameters. The Kolmogorov

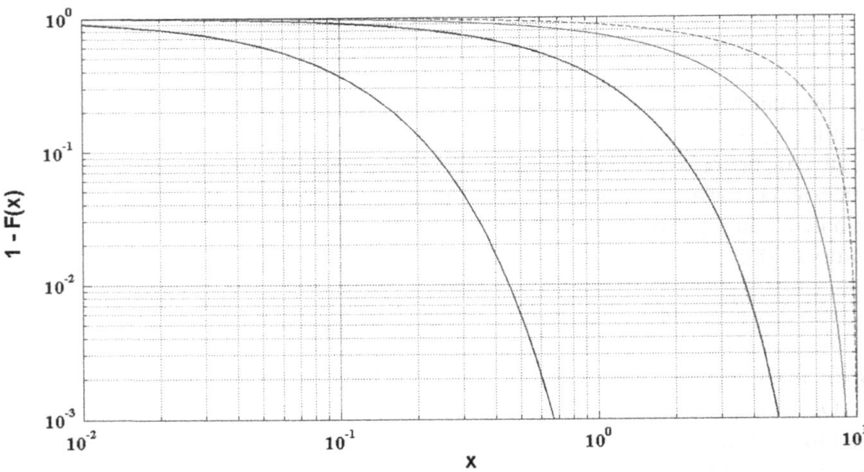

Fig. 2.1 GPD-tails. *Thick line* $\xi = -0.01; s = 0.1$; *intermediate line* $\xi = -0.10; s = 1.0$; *thin line* $\xi = -0.35; s = 3.5$ *dotted line* $\xi = -0.80; s = 8.0$; threshold h = 0; in all cases $M_{\max} = 10$

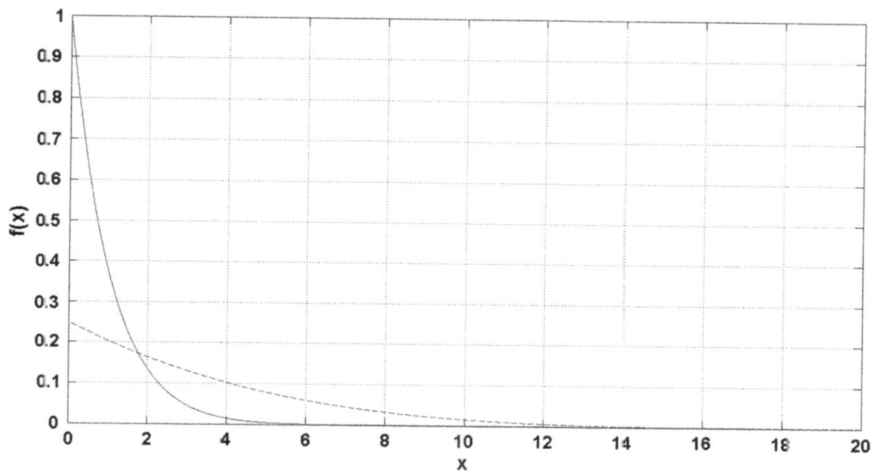

Fig. 2.2 GPD-densities. *Line* $\xi = -0.05$, $s = 1$. *Dotted line* $\xi = -0.20$, $s = 4$

test is a powerful statistical tool, but it needs a representative sample of sufficient size for fully reliable inference. This condition is not always fulfilled in practice, as we shall see below, since the limit theorem of EVT demands a sufficiently high threshold for its validity, and it is left less and less "peaks over threshold" for higher thresholds. This contradiction needs some compromise, sometimes resulting in small size of sample left for GPD-fitting.

The choice of threshold h should satisfy the following restrictions:

1. The threshold h should be high enough, so that the Limit Theorem of EVT mentioned above can be applied. The Kolmogorov distance *KD* between a fitted GPD and the actual sample DF should be small enough. The estimation of significance level for *KD* should take into account the fact that the two fitted GPD-parameters decrease the quantiles of the Kolmogorov distribution (see Pisarenko et al. 2008).
2. Sample size n of observations exceeding h should be large enough to provide reliable estimates of ξ, s and applicability of the Kolmogorov test. Our numerical experiments showed that it is necessary to have $n > 50 \div 80$ (however, in hard situations of deficit of data we were forced to use samples of size $n \cong 30$; of course, in such cases the reliability of statistical estimates is lowered).
3. If ξ is negative (positive), then parameter s (as it was shown in Pisarenko and Rodkin 2010) decreases (increases) with threshold h. Thus, it is reasonable to use thresholds providing decreasing (increasing) s-estimates. If $\xi = 0$ s-estimates should not significantly vary with h.

We use below the restrictions 1–3 to determine the range of h-values suitable for a proper estimation of parameters of the GPD-distribution fitting the tail of the studied sample.

2.2 Non-Stationarity of Natural Processes, the "Operational Time" Method

In this section we consider the problem of non-stationarity of catalogs of natural disasters and related losses. This aspect is very important for application of statistical methods to practical problems. Usually, catalogs consist of pairs (t, X), where t is time of an event (natural catastrophe like earthquake, tornado etc.) and X is information data about this event: coordinates, size, other characteristics. We are going to study events relating to a particular region, so coordinates just belong to this region. Thus, we can consider our catalogs consisting of pairs $(t_i,\ x_i)$, where t_i is time and x_i is size of the i-th event (earthquake magnitude, ground acceleration, number of fatalities, economic loss, etc.). The time occurrences t_i are modeled by a random point process. We shall use as a model the Poisson process with intensity λ (see Embrechts et al. 1997). As to the sizes x_i we assume, that they are random variables (independent of the Poisson process governing occurrences t_i), obeying to a certain (unknown) probability distribution with distribution function (DF) $F(x)$. The non-stationarity of the catalog can be caused by non-stationarity of the Poisson process (in this case the intensity is not constant, but

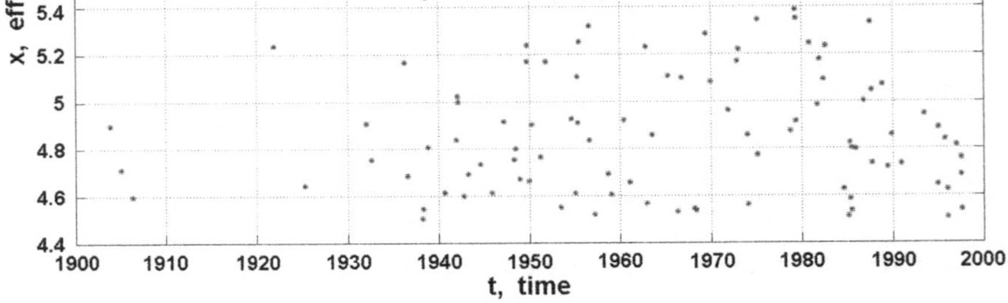

Fig. 2.3 Artificial sample ($n = 105$) of non-stationary Poisson process with the intensity $\lambda(t) = 0.375$, $1900 \le t \le 1940$; $\lambda(t) = 1.5$, $1940 < t \le 2000$. Exponential distribution of effect sizes: $F(x) = 1 - \exp[-2.0 \cdot (x - 4.5)]$, $x \ge 4.5$

depends on t: $\lambda(t)$) and by non-stationarity of the distribution $F(x)$, which results in dependence of distribution function $F(x)$ on time: $F(x, t)$. Of course, both these types of non-stationarity can occur simultaneously.

Figure 2.3 shows an artificial sample ($n = 105$) of non-stationary Poisson process with the intensity $\lambda(t)$:

$$\lambda(t) = \begin{matrix} 0.375\frac{1}{year}; & 1900 \leq t \leq 1940 \\ 1.5\frac{1}{year}; & 1940 < t \leq 2000 \end{matrix}. \tag{2.4}$$

$$F(x) = 1 - \exp[-2.0 \cdot (x - 4.5)], \quad x \geq 4.5 \text{ (exponential distribution)}.$$

Figure 2.4 shows a sample ($n = 300$) of non-stationary process with an intensity growing with time:

$$\lambda(t) = (t - 1900)/25 + 1; \quad 1900 \leq t \leq 2000 \tag{2.5}$$

Both figures give a general idea about behavior of intensity, but of course more rigorous statistical tools are needed for accurate estimation of $\lambda(t)$.

In order to test whether the occurrence times are generated by a stationary Poisson process (with constant $\lambda(t)$) we can use following known property of such processes: the conditional distribution of n time occurrences over time interval (0; T) under fixed n coincides with uniform distribution of n points on the interval (0; T). The uniformity of distribution can be tested by the standard Kolmogorov test. We take time occurrences on Fig. 2.3 and calculate the Kolmogorov distance $DK = \sqrt{n}\max|F_n(t) - t/T|$. Here $F_n(t)$ is sample DF of occurrences$F_n(t) = \#(t_i \leq t)/n$, and the maximum is taken over all values of t. We

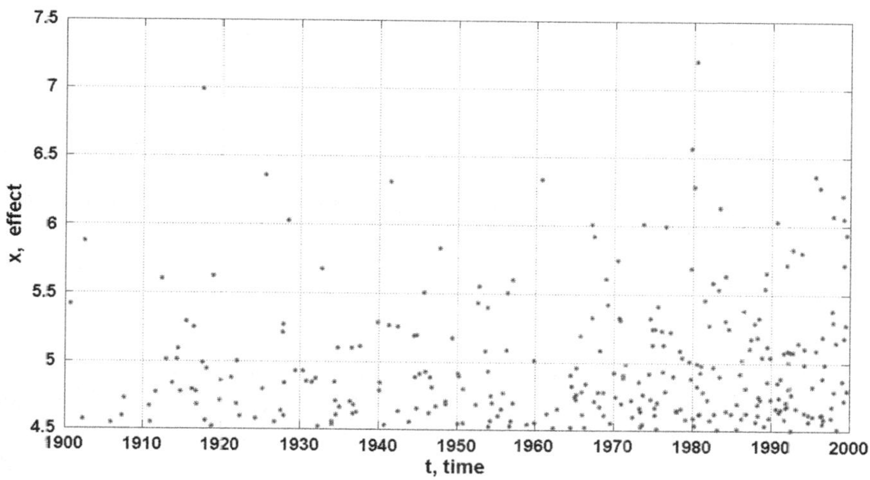

Fig. 2.4 Artificial sample ($n = 300$) of non-stationary process with an intensity growing with time: $\lambda(t) = (t - 1900)/25 + 1$; $1900 \leq t \leq 2000$. Exponential distribution of effect sizes: $F(x) = 1 - \exp[-2.0 \times (x - 4.5)]$, $x \geq 4.5$

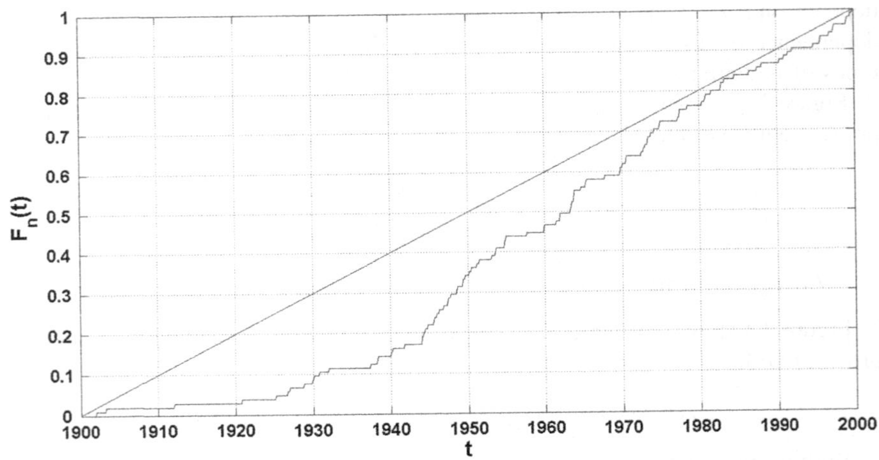

Fig. 2.5 Sample $F_n(t)$ of occurrence times t_j (Fig. 2.3) compared with the uniform DF (*diagonal line*)

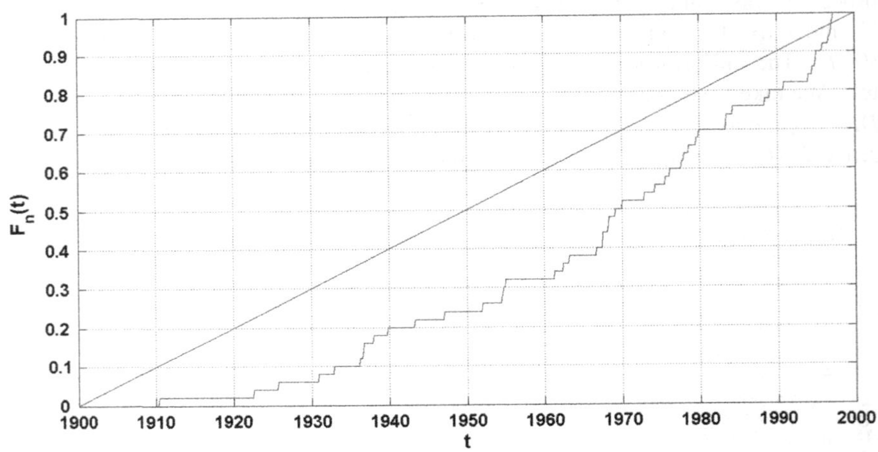

Fig. 2.6 Sample $F_n(t)$ of occurrence times t_j (Fig. 2.4) compared with the uniform DF (*diagonal line*)

get for Fig. 2.3 $DK = 2.79$ and for Fig. 2.4 $DK = 2.11$. On Figs. 2.5 and 2.6 we show corresponding sample $F_n(t)$ compared with the uniform DF. These figures correspond to small p-values: $p = 0.00083$ (Fig. 2.5) and $p = 0.023$ (Fig. 2.4). Thus, hypotheses of stationary should be rejected with very high confidence level.

On Fig. 2.5 one sees a more or less noticeable change of average slope somewhere near 1940 caused by jump of the intensity at $t = 1940$. In order to detect such intensity changes we suggest following procedure. We divide our

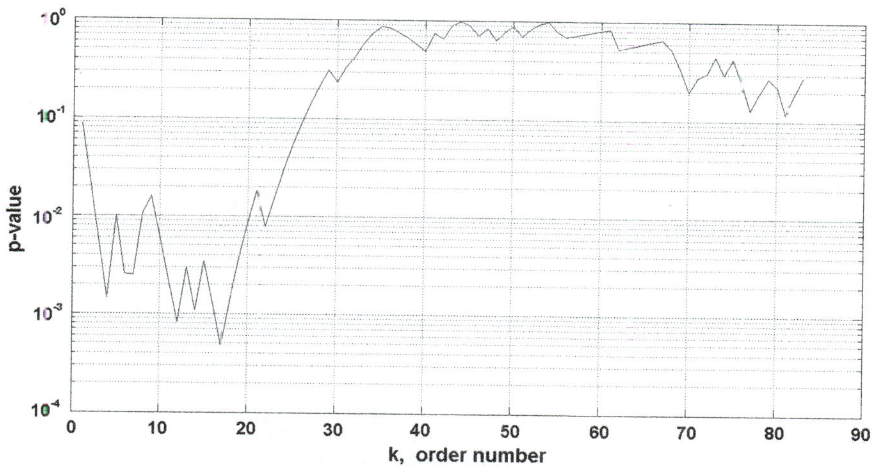

Fig. 2.7 *p*-values of K–S test: sample $(1 : k)$ versus sample $(k + 1 : n)$ as function of k. k is the order number of the kth event. Minimum point ($k = 18$) corresponds to the date 1940.3

catalog in two parts: sample S_t^1, containing occurrences $t_i \leq t$, and sample S_t^2, containing occurrences $t_i > t$. Now, we test the hypothesis H_0 that both samples are generated by one DF using the standard Kolmogorov–Smirnov test. Varying t we find t_0 giving the least p value. It corresponds to the most distinguishable samples $S_{t_0}^1$ and $S_{t_0}^2$. So, one can take point t_0 as an appropriate estimate of a sharp change of intensity. Figure 2.7 shows p-value as function of t. Minimum point t_0 corresponds to the date 1940.3.

An alternative way to judge about stationarity is allowed by looking at cumulative sums of time intervals between adjacent occurrences. If there is a tendency to increase (or decrease) intervals it often can be observed more or less clearly on cumulative curves. On Fig. 2.8 we see the cumulative sum of successive ordered time intervals between occurrences. A visible change of slope (marked by star) is observed at point 18, corresponding to the same time 1940.3 as minimum point at Fig. 2.7.

We have considered two examples of violation of stationarity represented by intensities (2.4)–(2.5). Of course, there are a lot of alternative ways of the violation. E.g. the intensity $\lambda(t)$ can be a polynomial of the 2nd or higher degree. Each practical situation needs its concrete study and appropriate statistical modeling. In our applications to real catalogs we restrict ourselves by intensity models (2.4)–(2.5) taking into account that non-stationarities in these catalogs can be satisfactorily modeled by (2.4)–(2.5), and the use of more sophisticated models are practically excluded by very limited size of available catalogs.

The non-stationarity of catalogs can be connected often with a range of event sizes. Say, seismic catalogs are less representative in lower ranges at the first half of the twentieth century. This fact can be explained by a low level of registration

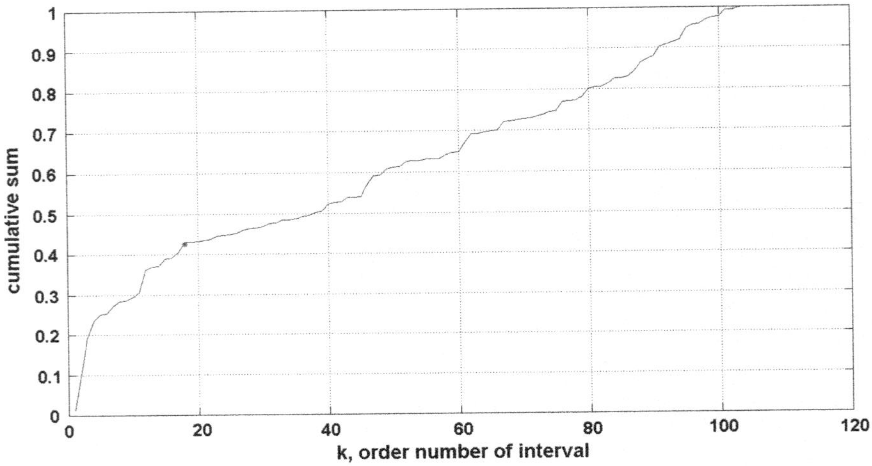

Fig. 2.8 Cumulative sum of successive time intervals between adjacent occurrences in sample Fig. 2.3. Point of slope break is marked by *star*

of small size eartquakes by existed at that time seismic networks. On the other hand, an evolution in preventive services measures can cause the essential decrease in fatalities and in loss values from the natural disasters. The latter tendency competes with a tendency of an loss increase due to the Earth's population growth and increasing value of the technosphere. The use of anti-seismic construction in some countries (e.g. in Japan) gives

an example of decrease of damages caused by earthquakes. Thus, it would be reasonable to look at intensity of events exceeding some lower threshold.

Let us consider an artificial example of seismic catalog. Suppose, seismic flow in the magnitude range $M \geq 4.5$ is formed by stationary Poisson time occurrences t with constant intensity $\lambda_0 = 5$ events per year. But the network registers earthquakes with random probability of registration $p(M, t)$:

$$p(M, t) = \begin{matrix} 0.4 \cdot M - 1.4; & 1900 \leq t < 1960; \\ 1; & 1960 \leq t \leq 2000. \end{matrix} \tag{2.6}$$

Thus, the intensity of evens registered in our catalog is $\lambda_0 \cdot p(M, t)$, and the stationarity is violated. Suppose further, that magnitude distribution is given by the Gutenberg-Richter law:

$$F(M) = 1 - \exp[-2.0 \cdot (M - 4.5)], \quad M \geq 4.5 \tag{2.7}$$

Thus, the stationarity is guaranteed only for events registered after $t = \mathit{1960}$. Figure 2.9 shows the magnitude-time diagram of our schematic example ($n = 299$). Figure 2.10 shows the intensity of three sub-catalogs corresponding to three lower magnitude thresholds: $h_1 = 5.0, n = 140$; $h_2 = 5.3, n = 87$; $h_3 = 6.0, n = 29$; intensities were smoothed by moving 40-year time interval.

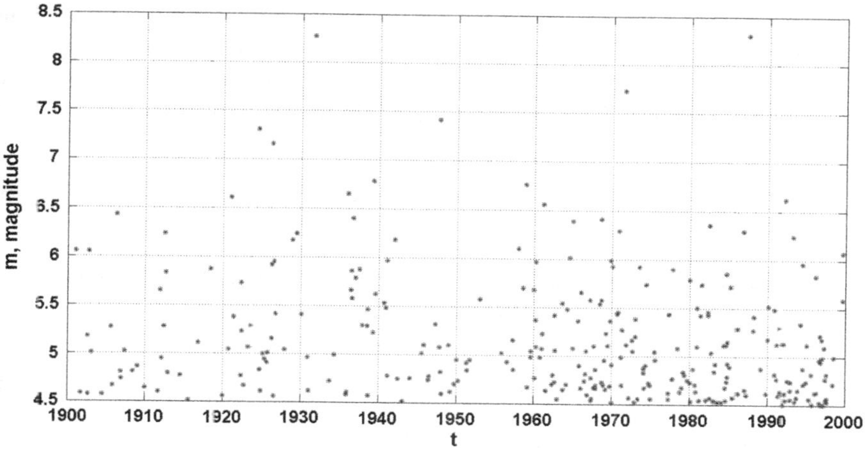

Fig. 2.9 Magnitude-time diagram of artificial example described by Eqs. (2.6) and (2.7), $n = 299$

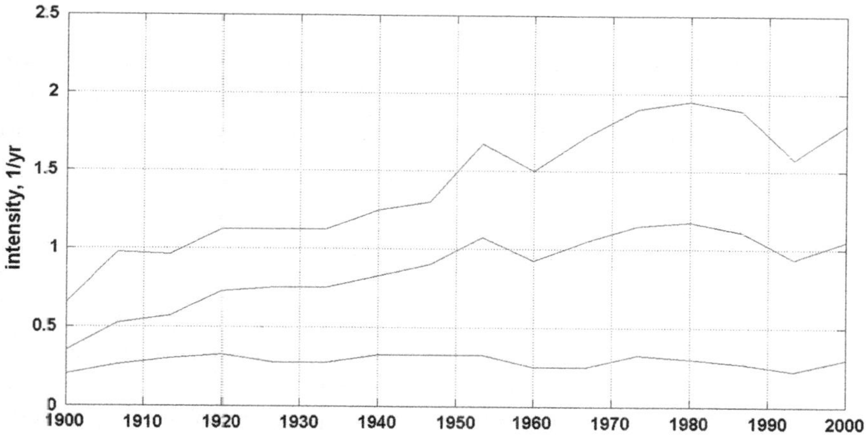

Fig. 2.10 Intensities of three sub-catalogs corresponding to three lower magnitude thresholds: $h_1 = 5.0$, $n = 140$ (*upper curve*); $h_2 = 5$, $n = 87$ (*middle curve*); $h_3 = 6.0$, $n = 29$ (*lower curve*), smoothed by 40-year window

We see that for thresholds $h_1 = 5.0$, $h_2 = 5.3$ intensities vary significantly, whereas for $h_3 = 6.0$ the intensity looks stable.

Suppose, the intensity λ of the point process $\xi(t)$ is not constant but vary with time as arbitrary positive function $\lambda(t)$. If we know $\lambda(t)$, we can transform time scale and pass to a new time $\tau(t)$ (sometimes, it is called "operational time"), so that the process $\xi(t(t))$ is stationary. We can apply to process $\xi(t(t))$ our statistical methods assuming stationarity (in particular, we can estimate desirable quantiles

$Q_q(t)$ for a future time interval τ, and then to return to the original time t. Important condition: the size distribution of sizes should not depend on time (i.e. stationarity of the distribution $F(x)$ is assumed). The needed time transformation is determined as follows. We take inverse function $G(.)$ with respect to $g(t)$

$$g(t) = \int_0^t \lambda(s)\,ds \tag{2.8}$$

Then the direct verification shows that point process $\xi[G(t)]$ is stationary with intensity one. Let us consider an example.

Suppose,

$$\lambda(s) = as + b; \quad a > 0 \tag{2.9}$$

Then

$$g(t) = at^2/2 + bt, \tag{2.10}$$

$$G(t) = \left[(b/a)^2 + 2t/a\right]^{1/2} - b/a \tag{2.11}$$

We affirm that the process $\eta(t) = \xi[\left((b/a)^2 + 2t/a\right)^{1/2} - b/a]$ is stationary with intensity one. Indeed, the mean number of occurrences of the process $\eta(t)$ on interval $[0, T]$ equals to mean number of occurrences of the process $\xi(t)$ on interval $[0, T_1]$, $T_1 = \sqrt{(b/a)^2 + 2T/a} - b/a$ which equals to the integral

$$\int_0^{T_1} \lambda(s)ds = \int_0^{T_1} (as + b)ds = \frac{aT_1^2}{2} + bT_1 = T.$$

The last identity just means that intensity of the process $\eta(t)$ equals unity.

Though the case examined above when the intensity of events' flow changes whereas the DF is stationary appears to be not quite natural it is a reasonable approach to the historical catalogs of strong earthquakes. The completeness of such catalogs depends frequently mainly on casual safety of information about ancient events whereas the DF is rather stationary for a long time because the slow change in technosphere of the ancient society.

Application of the operational time method to study of seismic regime can be found in Ogata (1989, 1993).

2.3 Parametrical Estimation of the Intensity $\lambda(t)$

The log-likelihood of a realization of event occurrences $(t_1, \ldots, t_n)$ in the time interval [0, T] is given by Vere-Jones (1995) and Ogata (1993):

$$\log L = \sum_{k=1}^{n} \log \lambda(t_k) - \int_0^T \lambda(s)ds \tag{2.12}$$

Suppose, that the dependence of the intensity $\lambda(t)$ on time can be modeled by some parametric function. For simplicity of exposition let us take linear function:

$$\lambda(t) = at + b. \tag{2.13}$$

Inserting (2.13) into (2.12) we get:

$$\log L(a, b) = \sum_{k=1}^{n} \log(at_k + b) - \frac{aT^2}{2} - bT \tag{2.14}$$

Now we can apply the full machinery of the likelihood methods for to derive estimators of parameters (a, b). The likelihood equations determining the estimators of maximum likelihood (MLE) are:

$$\frac{\partial L}{\partial a} = -\frac{T^2}{2} + \sum_{k=1}^{n} \frac{t_k}{at_k + b} = 0 \tag{2.15}$$

$$\frac{\partial L}{\partial b} = -T + \sum_{k=1}^{n} \frac{1}{at_k + b} = 0 \tag{2.16}$$

Using these equations we can express parameter a through b:

$$a = \frac{2(n - bT)}{T^2}. \tag{2.17}$$

Inserting (2.17) into (2.16), we get one equation with one unknown parameter b that should be solved numerically.

Using standard technique of the maximum likelihood estimation (Embrechts et al. 1997) one can derive standard deviations of MLE $\hat{a}$, $\hat{b}$:

$$std(\hat{a}) = \left[\frac{a}{T^2} \cdot \frac{w \log(1 + w)}{(1 + w/2)\log(1 + w) - w}\right]^{1/2}, \tag{2.18}$$

$$std(\hat{b}) = \left[\frac{b}{T} \cdot \frac{w(w/2 - 1) + \log(1 + w)}{(1 + w/2)\log(1 + w) - w}\right]^{1/2}, \tag{2.19}$$

where $w = aT/b$. In our example shown on Fig. 2.4 where $\lambda(1900 + t) = 0.04 \cdot t + 1,\ 0 \le t \le 100$, we got $\hat{a} = 0.034 \pm 0.0056$; $\hat{b} = 1.14 \pm 0.26$. Similarly one could use more sophisticated models for $\lambda(t)$.

Finishing discussion on the non-stationarity we consider a generalization of our problem, where the intensity depends both on t and on size x in a very general form given by function $\lambda(t,\ x)$ In this case the log-likelihood is:

$$\log L((t_1, x_1), \ldots (t_n, x_n)) = \sum_{k=1}^{n} \log \lambda(t_k, x_k) - \int_0^T ds \int_m^M \lambda(s, y) dy \qquad (2.20)$$

and the second integral represents the intensity of events at time s in the range $(m,\ M)$, where registration was effectuated. It should be remarked, that for fixed t the density $\lambda(t,\ x)$ is proportional to PDF $f(t, x) = \frac{\partial F(x,t)}{\partial x}$. Application of the exposed likelihood technique to numerous problems connected with seismic regime can be found in Vere-Jones (1995), Ogata (1993) and Lyubushin and Pisarenko (1994, 1998).

2.4 Annual Data

Sometimes statistics of the natural catastrophes are published in form of annual data. Now we are going to modify the exposed above method for such data.

Suppose, we have a list of sizes (economic losses, fatalities, etc.) representing yearly figures for N sequential years: $x_1, \ldots, x_N$ Our problem consists in statistical estimation of quantiles $Q_q(\tau)$ of maximum size in question for future τ years (now in contrast to the considered above situation τ is an entire number of years). We assume that the limit theorem (Theorem 3.4.13, Embrechts et al. 1997) is applicable to our data and there exists such sufficiently high threshold h that distribution of peaks over this threshold is well approximated by GPD. We denote parameters of this GPD by $\xi,\ s$, number of peaks by m (m depends on h), and the sample of peaks by

$$z_1, \ldots, z_m. \qquad (2.38)$$

We shall use ratio m/N as natural estimator of probability p exceeding h in future observations: $p \cong m/N$ According to our assumption the conditional distribution of x above h is:

$$GPD_h(x|\xi,\ s) = 1 - [1 + (\xi/s) \cdot (x - h)]^{-1/\xi}, \quad x \ge h. \qquad (2.39)$$

Parameters $\xi,\ s$ are estimated by the maximum likelihood method with use of sample (2.38), and the goodness of fit is tested by the Kolmogorov test taking into account the presence of two estimated parameters in the distribution function, as it was remarked earlier. Let us denote τ future sizes as $X_1, \ldots, X_\tau$. Here, the sizes are annual losses. We are interested in distribution of

$$M_\tau = \max(X_1, \ldots, X_\tau) \tag{2.40}$$

The random value X_1 (the annual loss) is less than h with probability $(1 - p)$. Under condition that $X_1 > h$ (which happens with probability p) its distribution is governed by GPD (2.39). Thus, for $x \geq h$ the distribution of X_1 equals to

$$(1 - p) + p \cdot GPD_h(x|\xi, s) \tag{2.41}$$

We do not study the distribution of X for range $X < h$ since this is of no (or very small) importance for distribution of M_τ. So, if $X < h$ we take $X \approx h$. Then distribution function of X_1, denoted as $F_1(x)$, is zero for $x < h$ and equals to (2.41) for arguments exceeding h. The distribution function $F_M(x)$ of the maximum M_τ equals to the τth degree of $F_1(x)$.

$$F_M(x) = \begin{matrix} 0; & x<h; \\ [(1-p) + p \cdot GPD_h(x|\xi, s)]^\tau; & x > h. \end{matrix} \tag{2.42}$$

We can find the q-level quantile $Q_q(\tau)$ of M_τ from equation:

$$F_M(x) = q \tag{2.43}$$

It follows from (2.42), that if $q > (1 - p)^\tau$, then

$$Q_q(\tau) = h + (s/\xi) \cdot \left[\left(\frac{1 - q^{1/\tau}}{p} \right)^{-\xi} - 1 \right] \tag{2.44}$$

For $q < (1 - p)^\tau$ the quantile $Q_q(t)$ is not defined. We could put it conditionally equal to h (it can be remarked that usually only quantiles of high level are of interest for risk problems).

If we take instead of $F_1(x)$ the exponential distribution with DF $1 - \exp(-\alpha \cdot (x - h))$, $x \geq h$ (we recall that the exponential distribution is the limit of GPD as $\xi \to 0$), then we get:

$$Q_q(\tau) = h - (1/\alpha) \cdot \log\left(\frac{1 - q^{1/\tau}}{p} \right) \tag{2.45}$$

References

Embrechts P, Klueppelberg C, Mikosch T (1997) Modelling extremal events. Springer, Berlin

Lyubushin AA, Pisarenko VF (1994) Research on seismic regime using linear model of intensity interaction point processes. Phys Solid Earth 29:1108–1113

Lyubushin AA, Pisarenko VF (1998) A new method for identifying seismicity periodicities. Volcanol Seismolog 20:73–89

Ogata Y (1989) Statistical model for standard seismicity and detection of anomalies by residual analysis. Tectonophysics 169:159–174

Ogata Y (1993) Fast likelihood computation of epidemic type aftershock-sequence model. Geophys Res Lett 20:2143–2146

Pisarenko VF, Rodkin MV (2010) Heavy-tailed distributions in disaster analysis. Springer, Dordrecht-Heidelberg-London-New York

Pisarenko VF, Sornette A, Sornette D, Rodkin MV (2008) New approach to the characterization of M_{max} and of the tail of the distribution of earthquake magnitudes. Pure Appl Geophys 65:847–888

Vere-Jones D (1995) Forecasting earthquakes and earthquake risk. Int J Forecast 11:503–538

Chapter 3
The Disaster Statistics for Various Natural Disasters

Abstract The application of statistical technique exposed in Chap. 2 to some concrete catalogs of natural processes and related losses are presented: global catalog of seismic moments; catalog of peak ground acceleration at five sites in Japan; catalog of victims of earthquakes-tsunamis, Japan; catalog of victims of floods, USA; catalog of economic losses from floods, USA; catalog of victims from tornadoes, USA; catalog of economic losses from hurricanes, USA. The Kolmogorov test is used as a powerful statistical tool for testing hypotheses on distribution under study. A modification of the Kolmogorov test is presented in the case of presence of estimated parameters in the hypothetical distribution function. Main statistical results are summarized in Tables 3.1, 4.1, 4.2 and 4.3.

Keywords Extreme value theory, EVT · Generalized Pareto distribution, GPD · "Peak-over-threshold" method · Intensity of point process · Non-stationarity of point process · Annual data

3.1 Earthquakes (Energy, Ground Acceleration)

3.1.1 *The Global Harvard Catalog of Scalar Seismic Moments*

We use the method described above for the Harvard catalog of seismic moments within the period from January 1, 1976 to October 31, 2012. We restrict the depth of epicenters to $h \leq 70$ km and magnitudes to $m_W \geq 6.25$ (or seismic moments $\geq 2.985 \times 10^{25}$ dyne-cm). Note that this time interval contains two recent gigantic earthquakes: December 26, 2004, $m_W = 9.0$ (Sumatra) and March 11, 2011, $m_W = 9.1$ (Japan). To eliminate aftershocks from the catalog, the space–time window suggested in Knopoff et al. (1982) was used. Scalar seismic moments M_s were converted into moment magnitude m_W by the relation

V. F. Pisarenko and M. V. Rodkin, *Statistical Analysis of Natural Disasters and Related Losses*, SpringerBriefs in Earth Sciences,
DOI: 10.1007/978-3-319-01454-8_3,

$$m_W = (2/3) \cdot (\log_{10}(M_S) - 16.1).$$

The number of main shocks that has been left after aftershock elimination was $n = 1073$.

In order to test the Harvard catalog for stationarity we plotted in Fig. 3.1 the intensity (main events). We see that the seismic flow can be considered as fairly stationary. The general view of event flow is shown on Fig. 3.2. It confirms our conclusion about stationarity, but still there is a slightly increased density of events in the upper right corner of Fig. 3.2. We have checked this suspicion and plotted on Fig. 3.3 smoothed by 15-year window intensity of events with $m_W \geq 8.0$. We see that indeed there is some distinct increasing of such events in the last 15–20 years. We shall recall this fact later as we estimate quantiles $Q_q(\tau)$.

As we mentioned above, time moments t_i of the stationary Poisson process are uniformly distributed on interval [0; T] for any fixed sample size n. On Fig. 3.4 we compare the empirical DF of normalized time moments s_i $\hat{F}_n(s)$ with uniform DF (diagonal of the square, $F(s) = s$). The normalization to the unite interval was effectuated by $s = (t - t_1)/(t_2 - t_1)$; t_1, t_2 are start and end of the catalog. The standard Kolmogorov test gives the Kolmogorov distance $D_n = \sqrt{n} \max |\hat{F}_n(s) - F(s)| = 1.127$ which corresponds to p value $= 0.16$ (probability to exceed D_n under condition that $\hat{F}_n(s)$ was generated by theoretical DF $F(s)$). Since this p-value (p–v) is more than 0.1, we formally have ground to accept the hypothesis of stationarity of t_i, although the p–v is close to the boundary of rejection 0.1.

Now we apply to the Harvard catalog the statistical analysis exposed above (GPD fitting). The graph of the sample tail $1 - F(x)$ is shown on Fig. 3.5. We see that the tail decays rather slowly, and the recent gigantic earthquakes form a deviation from the general run. We take a grid of thresholdsh_j for magnitudes m_W and fit both GPD

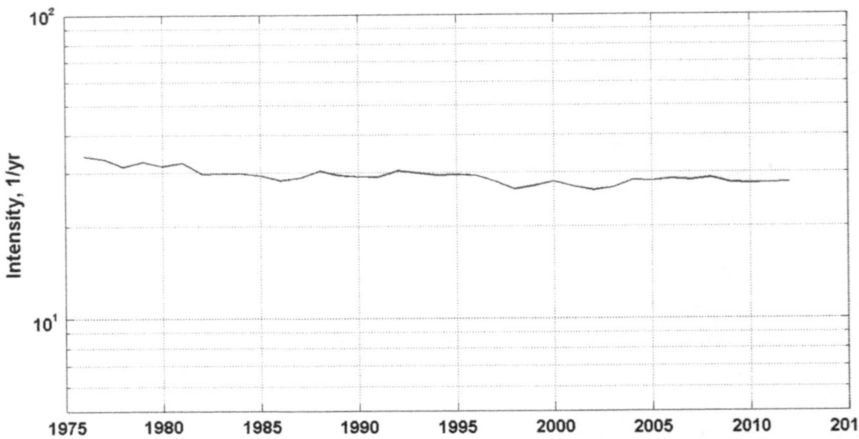

Fig. 3.1 Intensity (main events, $m_W \geq 6.25$) of the Harvard catalog smoothed by 10-year time window

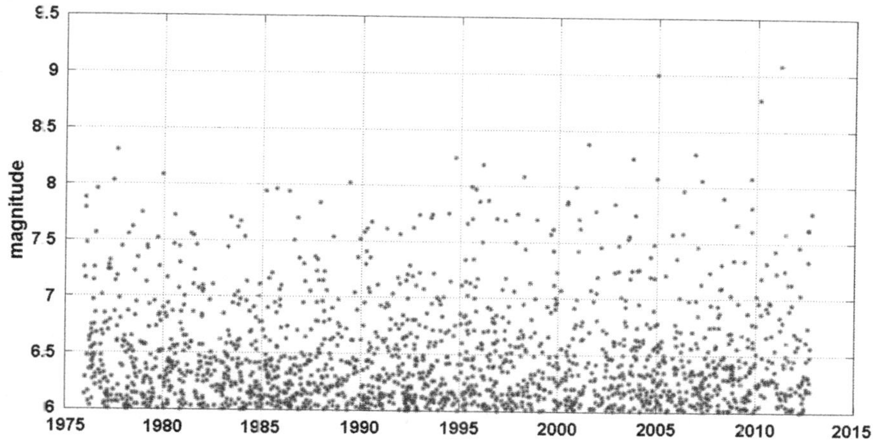

Fig. 3.2 The time–magnitude diagram of the Harvard catalog (main events, $m_W \geq 6.0$)

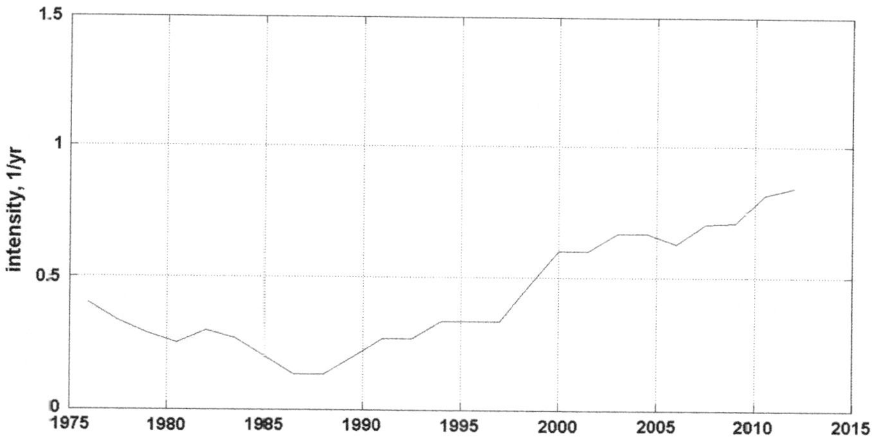

Fig. 3.3 Intensity (main events, $m_W \geq 8.0$) of the Harvard catalog smoothed by 15-year time window

and exponential distributions (ED) for each h_j. The goodness of fit was tested by the Kolmogorov test. The results are shown on Fig. 3.7. We see that minimum KD-distance is reached by GPD at $h = 6.8$ ($KD = 0.609$). We have estimated by the simulation method the goodness of this fit with parameters corresponding to this threshold ($\hat{\xi} = -0.163 \pm 0.046$; $\hat{s} = 0.540 \pm 0.055$) and found out that p–v= 0.15. So, we took the threshold $h = 6.8$ and corresponding estimates. On Fig. 3.6 the extreme part of the tail used for parameter estimation is shown along with fitted GPD-curve. We can remark that the three largest earthquakes

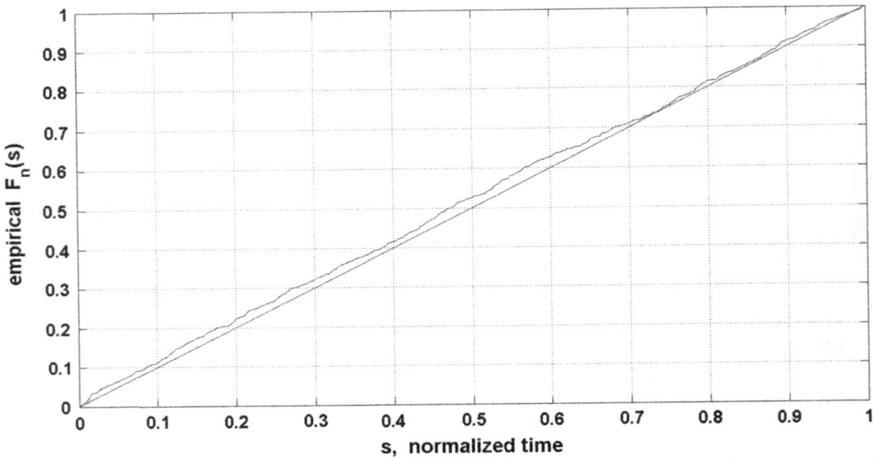

Fig. 3.4 Empirical DF of normalized time moments of events of the Harvard catalog (main events, $m_W \geq 6.25$)

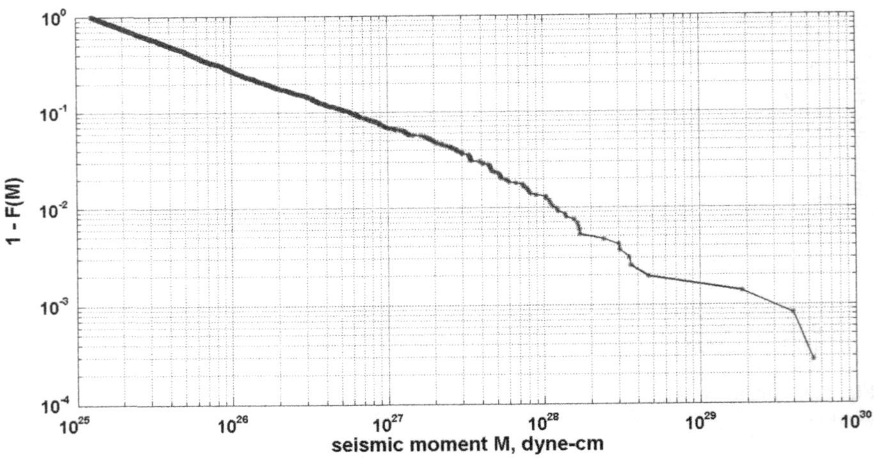

Fig. 3.5 Sample tail *1* − *F*(*x*) of the Harvard catalog: main events, seismic moments, $M_s \geq 1.26 \times 10^{25}$ dyne-cm ($m_W \geq 6.0$)

occurred in the last decade ($m_W = 8.8$, 27.2.2010; $m_W = 9.0$, 26.12.2004; $m_W = 9.1$, 11.3.2011) made the approximating GPD-curve to deviate from observations in the range $8.0 \leq m_W \leq 8.5$ (recall Figs. 3.2 and 3.3). One can say that this compromise is chosen in accordance with statistical rules used in our procedure of fitting and corresponds to the best goodness-of-fit possible in this situation.

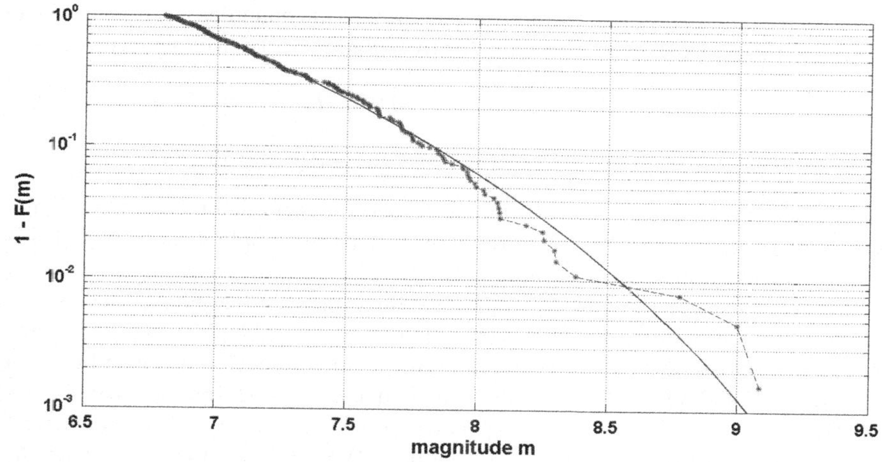

Fig. 3.6 Harvard catalog, main events, moment magnitudes. The extreme tail $1 - F(x)$ and approximating GPD-tail: $h = 6.8$; $\xi = -0.163$; $s = 0.540$; $n = 324$

The number of events exceeding this threshold was $n = 324$, which is sufficient for a reliable statistical estimation. Parameter M_{max} (right end-point of GPD) and 95 % quantile for future 10 years $Q_{0.95}(10)$ are:

$$M_{max} = h - \hat{s}/\hat{\xi} = 10.11; \; Q_{0.95}(10) = 9.13.$$

On Fig. 3.7 we see that besides of the best fitting point $h = 6.8$ there is one more point $h = 7.5$ whose KD is close to the best one. Let us compare parameters

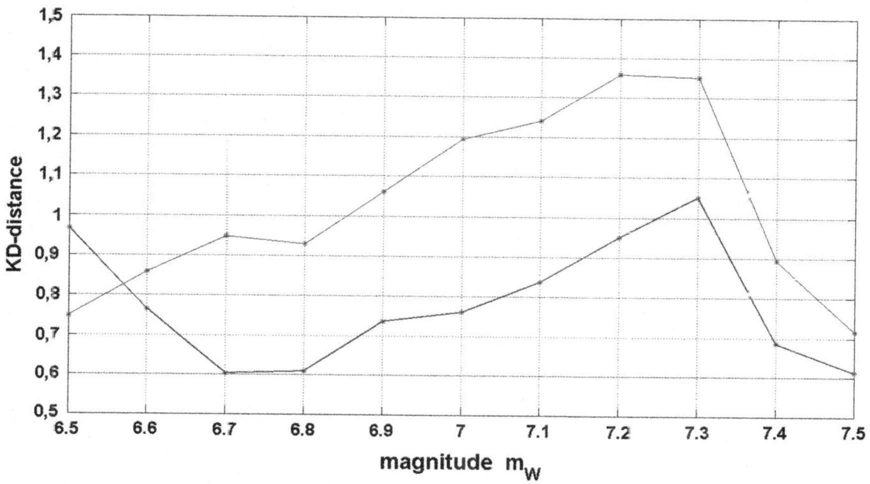

Fig. 3.7 Harvard catalog, main events. The KD-distances for a grid of magnitude-thresholds h_j. *Thick line*—GPD-fitting; *thin line*—ED-fitting

of these two thresholds and demonstrate stability of the quantiles as compared with traditional parameter M_{max}. We have for threshold $h = 7.5$ following estimates:

$$\hat{\xi} = -0.0597 \pm 0.104;\ \hat{s} = 0.347 \pm 0.074;\ KD = 0.612;\ n = 82.$$

$$M_{max} = h - \hat{s}/\hat{\xi} = 13.31;\ Q_{0.95}(10) = 9.27.$$

We see that quantile $Q_{0.95}(10)$ has increased (due to random errors) insignificantly by 0.14, whereas M_{max} has grown by 3.2! This example demonstrate clearly the instability of M_{max} as compared with $Q_q(\tau)$.

Figure 3.8 shows the GPD-quantiles for three different confidence levels $q = 0.90; 0.95; 0.99$ with parameters corresponding to threshold $h = 6.8$. It is interesting to compare these quantiles with corresponding quantiles estimated on time interval (1976–2006), that were published in the book (Pisarenko and Rodkin 2010), not containing the great Japan earthquake March 11, 2011. Figure 3.9 shows such comparison for 95 %-quantiles. We see that the later estimates (1976–2012) provide slightly larger quantiles. This is a reaction on the great Japan earthquake March 11, 2011. In contrast with the moderate change of quantile values, change in corresponding M_{max} values is much stronger that demonstrates again a weak stability of this parameter.

Note that even quantiles estimated for time interval before the occurrence of the great Tohoku earthquake (March 11, 2011, M9) show a quite high probability of occurrence of M9+ earthquakes. Besides, such events were not registered in Japan before both during the instrumental period of measurements and even among the historical earthquakes since 599 year (Usami 1979, 2002; Utsu 1979, 2002). Moreover, taking into account the considerable fragmentation of the lithosphere in

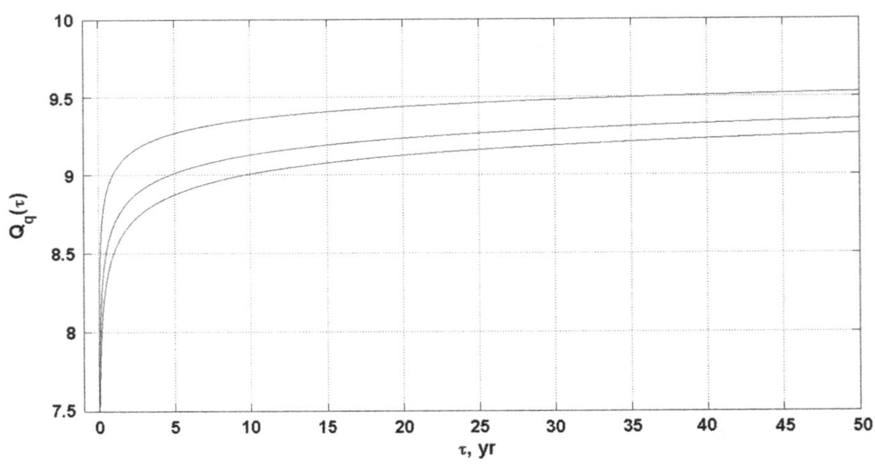

Fig. 3.8 Harvard catalog, main events. GPD-quantiles $Q_q(\tau)$ of seismic magnitudes for three different confidence levels $q = 0.90$ (*lower curve*); 0.95 (*middle curve*); 0.99 (*upper curve*) with parameters corresponding to the threshold $h = 6.8$

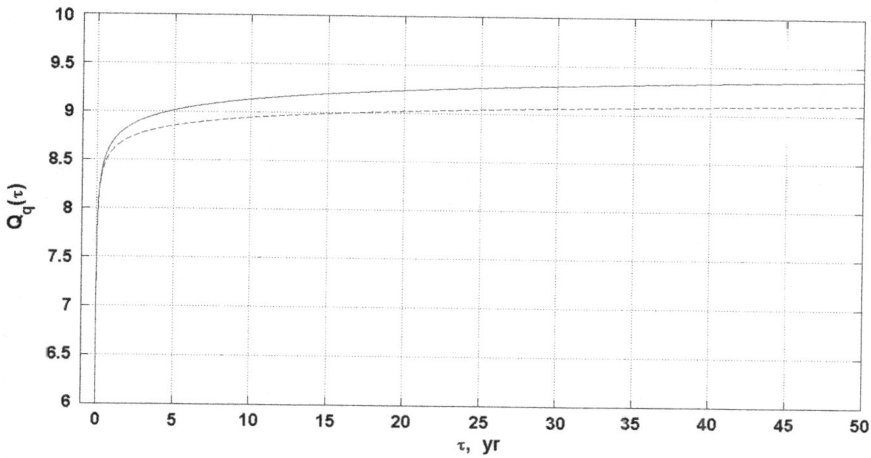

Fig. 3.9 Harvard catalog, main events. 95 % GPD-quantiles $Q_q(\tau)$ derived from data 1976–2012 (*thin line*) and quantiles $Q_q(\tau)$ for period 1977–2006 (*dotted line*)

the Japan region and the absence of extended (approaching 1000 km long) unified segments of the Benioff zone, the very possibility of occurrence of the M9+ earthquakes in this region was negated by many seismologists. The occurrence of the Great Tohoku 2011 earthquake has confirmed however that our statistical estimate of possibility of occurrence of such earthquakes Pisarenko et al. (2010); Pisarenko and Rodkin (2013) was quite correct.

3.1.2 Estimation of Maximum Peak Ground Acceleration

Although the main seismic parameters like *b*-slope, seismic activity rate and M_{max} can be of considerable interest, estimation of peak ground acceleration, A_{max}, is of more practical importance in designing structures and in seismic risk assessment. The earthquake hazard has been estimated in a variety of ways (Lamarre et al. 1992; Kijko and Sellevoll 1989, 1992; Campbell 1981; Cornell 1968). The characterization of the seismic hazard at a fixed site is usually done through the probability of non-exceeding various levels of ground acceleration in a certain number of years, i.e. through the probability distribution function of maximum peak acceleration, for a given time period *T*. An equivalent, but perhaps more convenient characteristic of seismic hazard is furnished by the quantiles of this distribution function (we recall, that quantile is inverse function of the distribution function). Seismic hazard analysis involves several unknown parameters and relations: seismic activity rate λ, parameters of magnitude-frequency law, attenuation model (for ground acceleration), source model, soil characteristics, a model for earthquake sequence. Thus, it is necessary to estimate these parameters and establish step by step the needed relations. Statistical and modeling uncertainties should be introduced at each of these steps. We are going to apply the statistical method exposed in Section II to evaluation of quantiles of distribution of peak ground acceleration.

We are interested in analysis of peak ground acceleration (PGA) A_{max} at a particular site. Suppose this site is located at epicentral distance R from source of earthquake of magnitude m. In the seismic hazard analysis there are a number of model relations giving approximate value of A_{max} as some function of (R, m), see e.g. Cornell (1968). In most cases these relations have the following general form:

$$\log_{10}(A_{max}) = a + b \cdot m - c \cdot \log(R + d), \tag{3.1}$$

where a, b, c, d are some non-negative coefficients, m is magnitude, R is epicentral distance. Numerous modifications of (3.1) are used, but all of them keep the general property: monotone increase with m and monotone decrease with R.

Let us consider flow of earthquakes in some space-magnitude window registered at a certain point. We denote magnitudes and epicentral distances to a fixed point as $(m_1, R_1), (m_2, R_2), \ldots, (m_n, R_n), \ldots$. We suppose that this flow is a stationary random process. Then any relation of type (3.1) or any arbitrary function $\Phi(m, R)$ will provide a stationary random process $\Phi(m_1, R_1)$, $\Phi(m_2, R_2), \ldots, \Phi(m_n, R_n)\ldots$. Thus, if we apply the relation (3.1) to the series $(m_1, R_1), (m_2, R_2), \ldots, (m_n, R_n), \ldots$. we can consider resulting sequence as a stationary random process. We call it *estimated acceleration*. The expression (3.1) differs from the true peak ground acceleration by a random term ε. We discuss this random error below. Our aim is to study statistical characteristics of the estimated acceleration with the statistical technique exposed in Chap. 2.

Detailed studies showed that the relation (3.1) is in some contradiction with empirical data in the near-field zone. In Mahdavian et al. (2005); Aptikaev (2009); Graizer and Kalkan (2011); Steinberg et al. (1993) it was shown that peak ground acceleration (PGA) practically does not depend on magnitude in a vicinity of the earthquake fault zone but depends on the type of the focal mechanism. The size of this zone D usually varies from a few km to 10 km depending on the magnitude and can be well scaled according to the empirical law

$$\partial \log D / \partial m \cong 0.34. \tag{3.2}$$

To meet the near-field zone data we shall use the Aptikaev's relation (Lamarre et al. 1992) where the near-field effects are taken into account:

$$\log_{10}(A_{max}) = \begin{cases} 2.76; & \rho \le 1; \\ 2.76 - 0.55 \cdot \log_{10}(\rho); & 1 \le \rho \le 10; \\ 3.50 - 1.29 \cdot \log_{10}(\rho); & 10 \le \rho; \end{cases} \tag{3.3}$$

where

A_{max} is estimated peak ground acceleration in cm/s^2;
$\rho = R \cdot 10^{-0.325(m-5)}$ magnitude scaled distance;
R is epicentral distance in km;
m is magnitude.

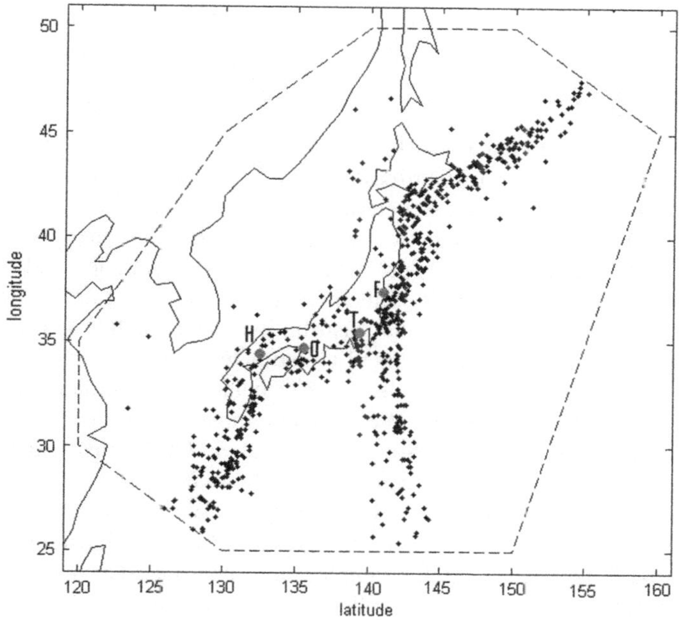

Fig. 3.10 Japan earthquakes, 1900–2005. Tokyo, Hiroshima, Osaka, and Fukushima (atomic power station Fukushima Daiichi) are marked as 'T', 'H', 'O', and 'F'

Thus, relations (3.1) and (3.3) give logarithm of estimated peak ground acceleration $\log_{10}(A_{max})$ from an earthquake with magnitude m at a site with epicentral distance R. The results obtained by means of both (3.3) and (3.1) are compared and discussed in Pisarenko and Rodkin (2013). In this paper we have derived quantiles of distribution of $\log_{10}(A_{max})$ for 4 points on the territory of Japan islands: Tokyo, Hiroshima, Osaka, and Fukushima (atomic power station Fukushima Daiichi). These points are shown on Fig. 3.10 where they are marked as 'T', 'H', 'O', and 'F'. We have used the earthquake catalog of the Japanese Meteorological Agency (JMA) over time period 1900–2005.

Tokyo, λ = 35.41; φ = 139.36;

We have applied the statistical technique exposed in Chap. 2 to the estimated accelerations calculated from equations (3.3). The Kolmogorov distance *KD* was used for to choose the most appropriate threshold value h providing the best fitting of *GPD* to the data:

$$KD = n_h^{1/2} \max |GPD_h(x \,|\, \hat{\xi}, \hat{s}) - F_{n_h}(x)|, \tag{3.4}$$

where $F_{n_h}(x)$ is sample stepwise distribution function generated by observations $(x_1 \leq \ldots \leq x_{n_h})$ exceeding threshold h:

$$F_{n_h}(x) = \begin{cases} 0; & x \le x_1; \\ r/n_h; & x_r < x \le x_{r+1}; \quad 1 < r < n_h; \\ 1; & x > x_{n_h}. \end{cases}$$

Since we use a theoretical GPD with parameters fitted to the data, we cannot use the standard Kolmogorov distribution tables to find the significance level of the observed *KD*. Instead, in order to determine the significance level of a given *KD*-distance (3.4), we used a numerically calculated distribution of *KD*-distances measured in a simulation procedure with 10,000 GPD-samples and parameters individually fitted to each sample. This method was suggested in Stephens (1974) for the Gaussian and the exponential distributions. We use the Kolmogorov distance to test the GPD distribution fitted to estimated accelerations.

The histogram of estimated accelerations calculated in accordance with (3.3) is shown on Fig. 3.11 (upper figure). We see that a monotone decreasing starts somewhere near $\log_{10}(A_{max}) \sim -0.2$. Since the theoretical GPD-density monotonically decreases, we have to restrict our analysis by thresholds $h > -0.2$. The *KD*-distance as function of h is shown on Fig. 3.11, lower figure. We see that the lowest $KD = 0.582$ (the best fitting) corresponds to the threshold $h = 0.5$. Its significance level (p–value) equals to 0.72, so that the sample can be considered as belonging to GPD distribution (the testing would reject this distribution in the case of very small p–values, say, $p < 0.10$). For this threshold there are $n_h = 279$ observations exceeding this threshold. We got following estimates of unknown parameters:

$$\hat{\xi} = -0.23 \pm 0.05; \quad \hat{s} = 0.52 \pm 0.06; \quad Q_{0.90}(30) = 2.3. \qquad (3.5)$$

Maximum of logarithmic estimated accelerations was 2.76 (it cannot be more because of restriction of estimated acceleration as it is given in equation (3.3)).

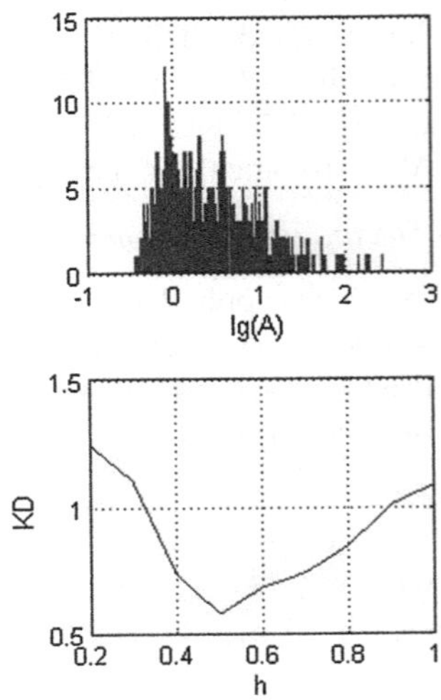

Fig. 3.11 Tokio. The histogram of estimated accelerations (3.3) (*upper figure*). The *KD*-distance as function of *h* (*lower figure*)

***Hiroshima,* λ = 34.39; φ = 132.46;**

The histogram of estimated accelerations calculated in accordance with (3.3) is shown on Fig. 3.12, upper figure. We see that a monotone decreasing starts somewhere near $\log_{10}(A_{max}) \sim 0.1$. We restrict our analysis by thresholds h in the interval (0.1; 0.9). The *KD*-distance as function of h is shown on Fig. 3.12, lower figure. We see that there are 3 thresholds $h = 0.6; 0.7; 0.8$ with *KD* close to 0.6. We prefer to take $h = 0.6$ since sample size for this threshold ($n_h = 118$) is larger than others ($n_h = 95$; $n_h = 74$). Its significance level equals to 0.593, so that the sample can be considered as belonging to GPD distribution. We got following estimates of unknown parameters:

$$\hat{\xi} = -0.21 \pm 0.07; \quad \hat{s} = 0.46 \pm 0.08; \quad Q_{0.90}(30) = 2.1. \tag{3.6}$$

Maximum of logarithmic estimated accelerations was 2.76 (it cannot be more because of restriction of estimated acceleration in Eq. (3.3)).

***Osaka,* λ = 34.69; φ = 135.50;**

The histogram of estimated accelerations calculated in accordance with (3.3) is shown on Fig. 3.13, upper figure. We see that a monotone decreasing starts somewhere near $\log_{10}(A_{max}) \sim 0.1$.We restrict our analysis by thresholds h in the interval (0.1; 0.6). The *KD*-distance as function of h is shown on Fig. 3.13, lower figure. We see that the best fitting corresponds to the threshold $h = 0.1$ with $KD = 0.62$. Its p–value equals to 0.63, so that the sample can be considered as belonging to GPD distribution. We got the following estimates of unknown parameters:

$$\hat{\xi} = -0.19 \pm 0.04; \quad \hat{s} = 0.50 \pm 0.04; \quad Q_{0.90}(30) = 2.1. \tag{3.7}$$

Maximum of logarithmic estimated acceleration was 2.76 (as above it cannot be more because of restriction of estimated acceleration in equation (3.3)).

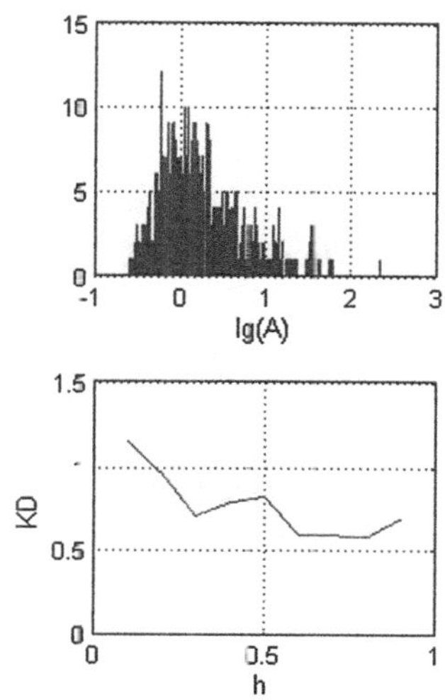

Fig. 3.12 Hiroshima. The histogram of estimated accelerations (3.3) (*upper figure*). The *KD*-distance as function of *h* (*lower figure*)

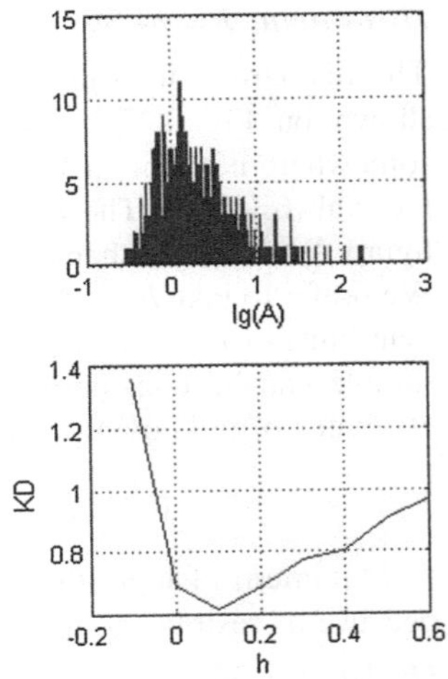

Fig. 3.13 Osaka. The histogram of estimated accelerations (3.3) (*upper figure*). The *KD*-distance as function of *h* (*lower figure*)

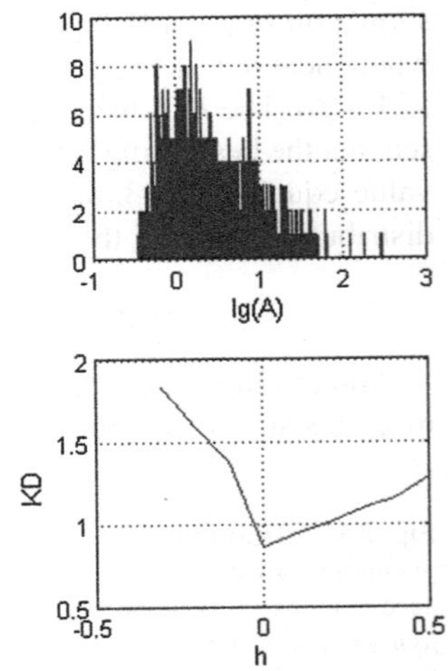

Fig. 3.14 Fukushima. The histogram of estimated accelerations (3.3) (*upper figure*). The *KD*-distance as function of *h* (*lower figure*)

Fukushima Daiichi, $\boldsymbol{\lambda = 37.4214}$***;*** $\boldsymbol{\varphi = 141.0325}$;

The histogram of estimated accelerations calculated in accordance with (3.3) is shown on Fig. 3.14, upper figure. We see that a monotone decreasing starts somewhere near $\log_{10}(A_{max}) \sim 0.2$.We restrict our analysis by thresholds h in the interval (0.0; 0.5). The *KD*-distance as function of h is shown on Fig. 3.14, lower figure. We see that the best fitting corresponds to the threshold $h = 0$ with $KD = 0.87$. Its p-value equals to 0.13, so that the sample still can be considered as belonging to GPD distribution. We got following estimates of unknown parameters:

Table 3.1 Statistical estimates of *GPD* parameters fitted to the estimated acceleration data

	h	n_h	ξ	$h - s/\xi$	$Q_{0.90}$ (30)	p-value of *GPD*
Tokyo	0.5	279	-0.23 ± 0.05	2.76	2.3	0.72
Hiroshima	0.6	118	-0.21 ± 0.07	2.79	2.1	0.69
Osaka	0.1	462	-0.19 ± 0.04	2.73	2.1	0.63
Fukushima	0	544	-0.29 ± 0.03	2.62	2.3	0.13

$$\hat{\xi} = -0.29 \pm 0.03; \quad \hat{s} = 0.76 \pm 0.06; \quad Q_{0.90}(30) = 2.3. \tag{3.8}$$

Maximum of logarithmic estimated accelerations was 2.63.

The results of estimation of the parameters of GPD distribution fitted to the estimated acceleration data for the mentioned four sites are summarized in Table 3.1. In the fifth column the maximum possible values of the estimated acceleration are shown calculated by formula $\log(A_{max}) = h - s/\xi$. $Q_{0.90}(30)$ is the quantile of level $q = 0.90$ for the maximum estimated acceleration in a future time interval of 30 years.

3.1.3 The Accounting for Inaccuracy of the Estimated Acceleration

The estimated acceleration (3.3) differs from the true acceleration by a random value ε. We assume that

$$\varepsilon = \varepsilon_1 + \varepsilon_2,$$

where ε_1, ε_2 are independent random errors; ε_1 refers to inaccuracy of the used relations (3.3) and ε_2 characterizes the influence of the seismic source mechanism on the ground acceleration. In accordance with Aptikaev (2009) the random error of the relation (3.3) has standard deviation std $(\varepsilon_1) = 0.18$. The distribution of ε_1 is not critical: we compared on several artificial examples the Gauss distribution and the uniform distribution and found no essential differences in the estimates of quantiles $Q_q(\tau)$. So, we accept for ε_1 the Gauss distribution. In order to evaluate std(ε_2). we suppose that all sources in Japan territory can be classified into three types with following relative frequencies:

$$\begin{aligned} \text{normal fault} \quad & \sim 15\,\%; \\ \text{strike} - \text{slip} \quad & \sim 20\,\%; \\ \text{inverse fault (thrust)} \quad & \sim 65\,\%. \end{aligned} \tag{3.9}$$

These relative frequencies are taken from the regional earthquakes focal mechanism data Zlobin and Polets (2012), and they reflect the predominance of the compression tectonic forces in the Japan region. Following Aptikaev (2009) we assume further that these source types produce correspondingly in the epicentral zone the following mean peak ground accelerations (PGA):

$$\begin{aligned} \log_{10}(A) &= 2.65; \\ \log_{10}(A) &= 2.76; \\ \log_{10}(A) &= 2.95. \end{aligned} \tag{3.10}$$

We have used here the mean PGA value $\log_{10}(A) = 2.76$ valid for a totality of earthquakes with different types of focal mechanisms instead of that typical of strike-slip events and equal to 2.80 Aptikaev (2009). It gives us a possibility to use a simpler scheme of taking into account the difference of mean PGA values in the cases of different focal mechanisms. If we knew the source mechanism for each earthquake in our catalog we could take this information into account, but since it is unknown for the JMA catalog (at least for the first half of this catalog), we have to model the influence of the source mechanism by an additional random term ε_2. The mean value of random variable taking values (3.10) with probabilities (3.9) is ~ 2.76 and standard deviation is 0.15. Thus, we can accept that std of ε_2 is 0.15. We suppose that the distribution of ε_2 is the Gaussian as well. Then the error $(\varepsilon_1 + \varepsilon_2)$ has standard deviation 0.23. Thus, we can assume that the maximum estimated acceleration analyzed in the previous section differs from the true maximum ground acceleration by a random Gaussian error with zero mean and $std = 0.23$. So, we have to take into account the influence of this random error on the quantile $Q_q(\tau)$. We have done it by a simulation procedure, adding a random Gaussian rv with $std = 0.234$ to the GPD-random variable with estimated parameters (see Table 3.1) and repeating this operation 10,000 times. Figure 3.15 shows the quantiles $Q_q(\tau)$. both with error term $\varepsilon = (\varepsilon_1 + \varepsilon_2)$ (heavy curves) and without it (light curve) for all four points under analysis. We see that the accounting for errors is practically reduced to an increase of the undisturbed quantile by one std of the error.

The quantile $Q_q(\tau)$ is a final result of our statistical technique. It is a robust and meaningful characteristic of the seismic hazard.

It should be remarked that the estimation of the peak ground acceleration depends in a large extent on the used relation of type (3.1) or (3.3) connecting the log-acceleration with magnitude and distance. We have used the relation (3.3) due to Aptikaev (2009) that does not take into account the regional or local features connected with soil properties. Of course, relation taking into account regional and geological pecularities of the site would be preferable. So, our results exposed above might be considered as preliminary estimation of real acceleration and illustration of our statistical method for this problem.

3.2 Earthquakes, Tsunami (Victims)

In this section we shall use revised and extended version of the catalog of earthquake victims in Japan composed by Utsu (1979, 2002). The catalog includes as well victims of tsunami, caused by earthquakes, and covers time period 1900–2012. It contains two types of data: fatalities and the injured.

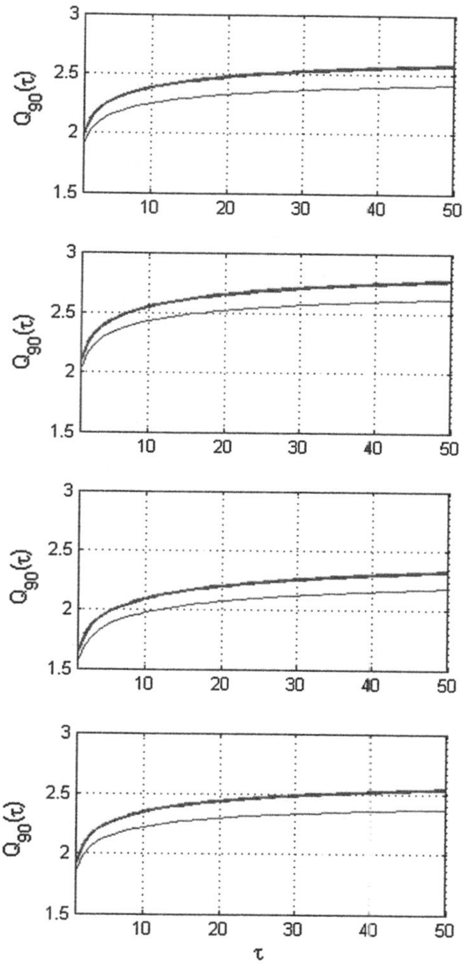

Fig. 3.15 GPD-quantiles $Q_{0.90}(\tau)$ of estimated acceleration both with error term $\varepsilon = (\varepsilon_1 + \varepsilon_2)$ (*heavy curves*) and without it (*light curves*) for all four points under analysis. From *top* to *bottom*: Tokyo, Hiroshima, Osaka, Fukushima

(a) ***Fatalities***

The catalog contains fatality data of 73 earthquakes. Figure 3.16 shows general view of these data. The maximum fatality (142,807) occurred at 1923, Tokio. Such high number can be explained by gigantic fires that struck old houses and structures existed at that time. So, if one is interested only in possibilities of similar disasters in the future this observation, perhaps, could be eliminated from the sample, but we left it nevertheless, counting for the robustness of our method. Looking at Fig. 3.16 we see that the intensity of events visually slightly decreases with time, although this effect is not very strong. Perhaps, this decrease can be explained by more safe modern structure and more effective preventive measures. It should be noted that this effect competes with the expected increase in a number of fatalities because of the natural growth of population.

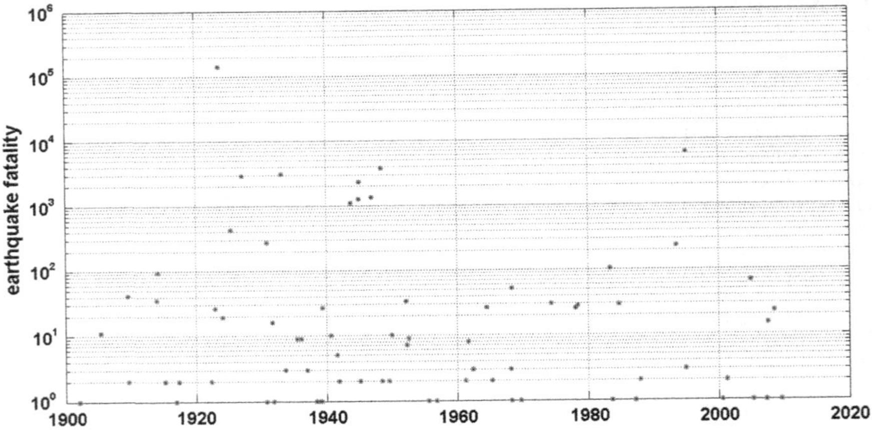

Fig. 3.16 The fatality events of the catalog of earthquake victims (Japan) composed by Utsu, 1900–2012

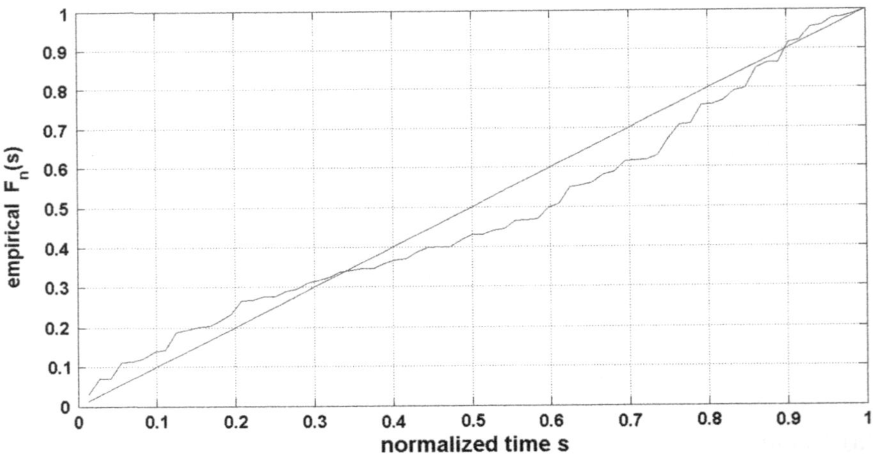

Fig. 3.17 Catalog of earthquake victims, Japan, 1900–2012. The empirical distribution function $F_n(s)$) of normalized occurrence times s_j compared with the uniform distribution function (*diagonal line*)

As it was mentioned above for the stationary Poisson process time moments t_i are uniformly distributed on interval $[0;\ T]$. On Fig. 3.17 we compare the empirical DF of normalized time moments s_i $\hat{F}_n(s)$ with uniform DF ($F(s) = s;$ $0 \leq s \leq 1$). The standard Kolmogorov test gives the Kolmogorov distance $D_n = \sqrt{n}\ max\ |\ \hat{F}_n(s) - F(s)| = 1.004$ which corresponds to p-value $= 0.23$ (probability to exceed D_n under condition that $\hat{F}_n(s)$ was generated by theoretical

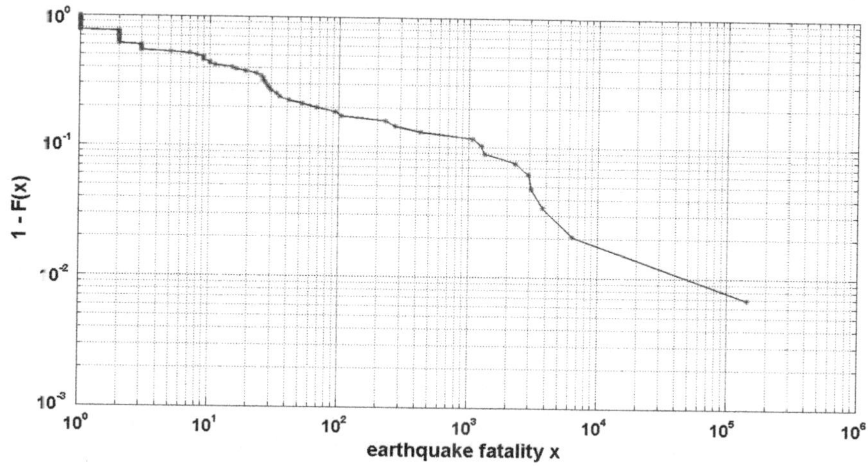

Fig. 3.18 Catalog of earthquake victims, Japan, 1900–2012. The tail graph of fatalities

DF $F(s)$. Since this p-value (p–v) is not sufficiently small (more than 0.1), we can accept the hypothesis of stationarity of t_i.

Now we consider the fatalities figures. The tail graph is shown on Fig. 3.18. We see that on the whole the tail $1- F(x)$ has power-like behavior with appreciable irregularities. So, it is reasonable to take logarithms and to analyze $\log(x)$. The concentration of victims in the extreme range of the catalog is very high: it turns out that 10 % of the most disastrous events are responsible for 98 % of the total number of perished. Even if we exclude maximum fatality case (142, 807 occurred at 1923, Tokio) 10 % of the most disastrous events would make up 83 % of the total number of perished.

On Fig. 3.20 the cumulative sums of log-fatalities are shown. In spite of some local fluctuations the trend on the whole looks like a linear function. There are some deviations, but they have no definite tendency. So, we accept the hypothesis of stationarity of this catalog.

Now we apply to our catalog the statistical analysis exposed above. We take a grid of thresholds h_j for $\log(x)$ and fit both GPD and ED for each h_j. The goodness of fit is measured by the Kolmogorov distance. The results are shown on Fig. 3.21. We see that minimum KD-distance 0.665 is reached for GPD-approach at $h = 0.45$. The GPD fitting includes estimation of two parameters (ξ, s), whereas the standard Kolmogorov testing assumes no unknown parameter. For this reason we cannot use directly the standard p–v from tables of the Kolmogorov distribution. In order to calculate p–v we used the simulation method as we mentioned above. The best fit of GPD-distribution for $h = 0.45$ ($\xi = -0.260 \pm 0.111$; $s = 1.657 \pm 0.430$) provides p–$v = 0.433$, which allows to accept the GPD. Figure 3.19 shows the extreme part of the tail used for parameter estimation along

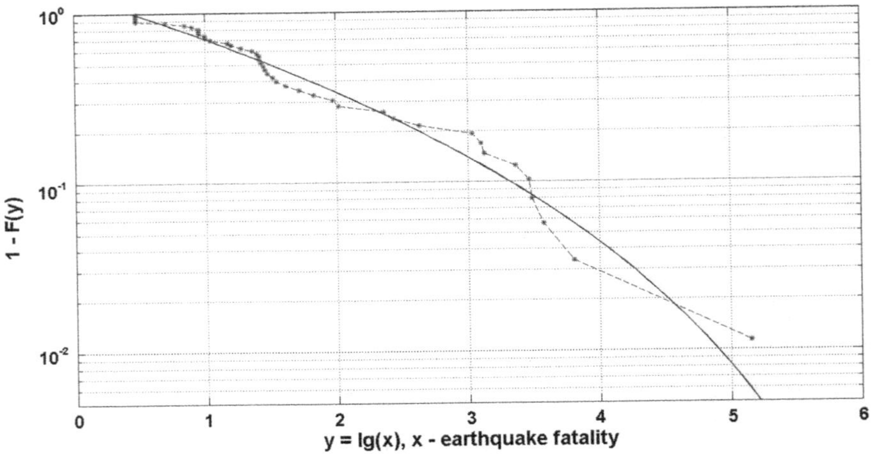

Fig. 3.19 Catalog of earthquake victims, Japan, 1900–2012. The extreme tail $1 - F(y)$ ($y = \log(x)$, x—number of dead) and approximating GPD-tail: $h = 0.45$; $\xi = -0.260$; $s = 1.657$; $n = 44$

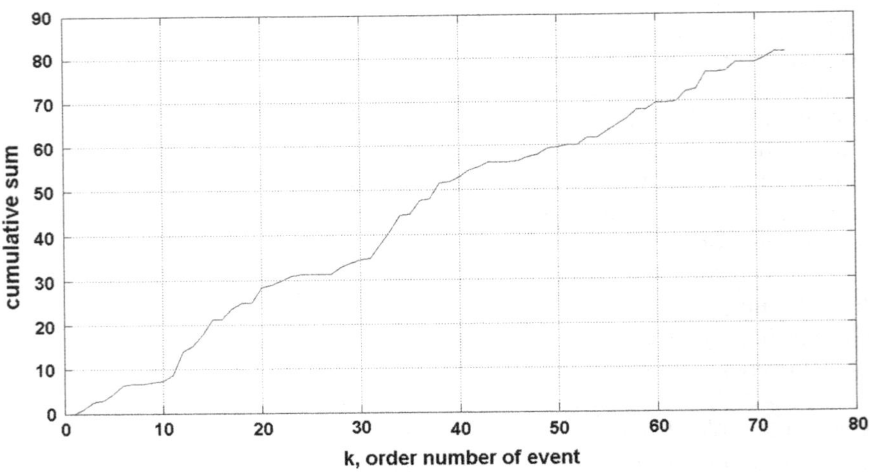

Fig. 3.20 Catalog of earthquake victims, Japan, 1900–2012. The cumulative sums of log-fatalities

with fitted GPD-curve. We see that observations exhibit some irregular oscillations around the approximating GPD-curve, but on the whole the fitting is satisfactory.

Now we are able to calculate the quantiles $Q_q(\tau)$ which are the final goal of our estimation. Figure 3.22 shows the GPD-quantiles for 3 different confidence levels $q = 0.90$; 0.95; 0.99. It should be remarked that for very small τ the quantiles

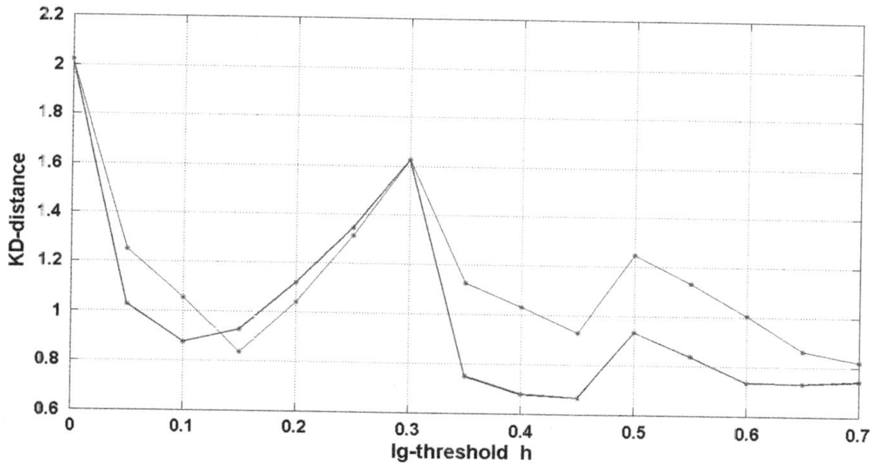

Fig. 3.21 Catalog of earthquake victims, Japan, 1900–2012. The KD-distances for a grid of log-thresholds h_j. *Thick line*—GPD-fitting; *thin line*—ED-fitting

takes zero values and have small jumps equal to $exp(-\lambda\tau)$ which is the probability that there is no event on the interval $[0; \tau]$.

It is interesting to calculate "maximum possible size", i.e. the rightmost limit M_{max} of the GPD for $\log(x)$:

$M_{max} = h - s/\xi = 6.83$ (this corresponds to 6,800,000 fatalities).

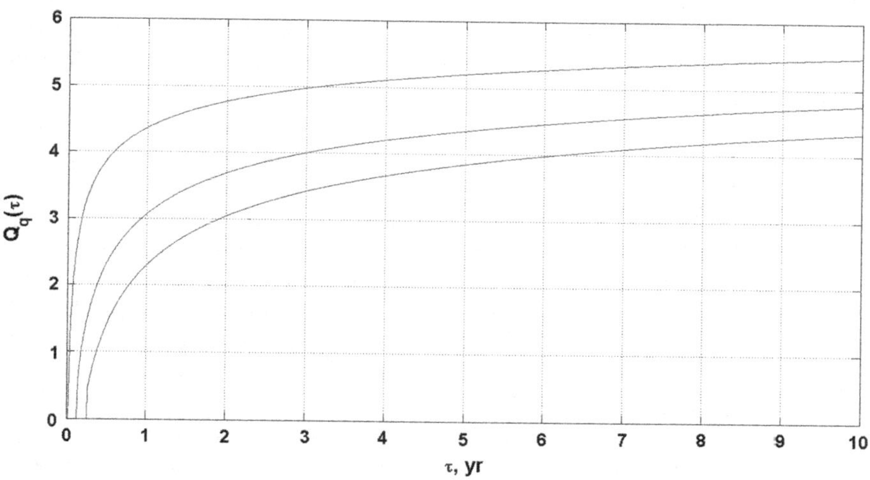

Fig. 3.22 Catalog of earthquake victims, Japan, 1900–2012. The GPD-quantiles for three different confidence levels $q = 0.90$ (*lower curve*); 0.95 (*middle curve*); 0.99 (*upper curve*)

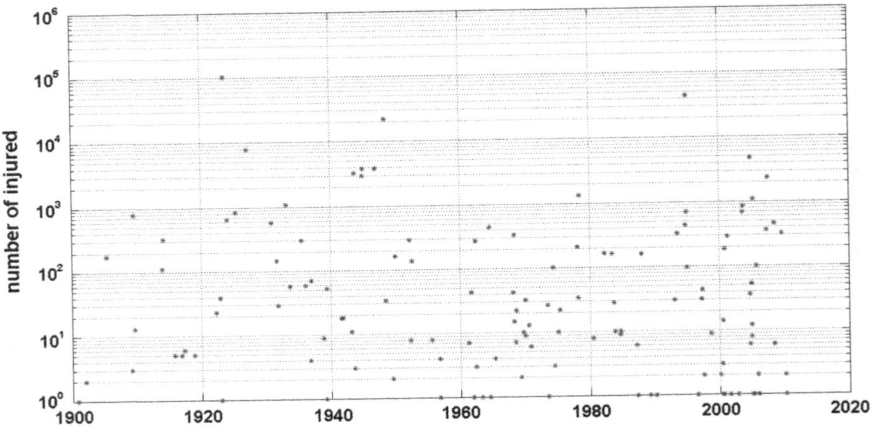

Fig. 3.23 The injured by earthquakes. Catalog of earthquake victims (Japan), 1900–2012

Such gigantic figure appears to be unreal, it hardly can be used as a useful statistical characteristic of real fatalities and has little practical value. In the same time the quantile $Q_{0.95}(10) = 5.66$ (460,000) looks quite realistic. This comparison shows once more stability of the quantiles $Q_q(\tau)$ with respect to the "maximum possible size" parameter.

(b) *The injured*

The catalog contains numbers of injured in 131 earthquakes. Figure 3.23 shows these numbers in time. The maximum number (103,733) occurred at 1923, Tokio. As we mentioned above such high number of dead and injured can be explained by gigantic fires. Looking at Fig. 3.23 we see that the intensity of events visually slightly increases with time, although this effect is not very strong. Perhaps, this increase can be explained by more attentive registration of events with minor number of injured.

On Fig. 3.24 we compare the empirical DF of normalized time moments s_i $\hat{F}_n(s)$ with uniform DF. We see that there is a certain down deviation of the empirical DF from the diagonal, which testifies that the visual effect of an intensity increase is real. The Kolmogorov distance $D_n = 2.048$ corresponds to p–$v = 0.0005$ which makes us to reject the hypothesis of stationarity of the catalog. Looking at Fig. 3.23 we can suspect that the non-stationarity is caused by weaker events. In order to check this suspicion we tried several lower thresholds. The resulting intensities smoothed by 15-year time window are shown on Fig. 3.25. We see that the evident non-stationary intensity of the original catalog ($h = 0$) decreases with growing h and practically vanishes at $h = 400$. This threshold seems high (it is left only 23 observations above this threshold), but, fortunately, just this threshold provides the best GPD-fitting, as we shall see below. The empirical DF of normalized event times for $h = 400$ is shown on Fig. 3.26.

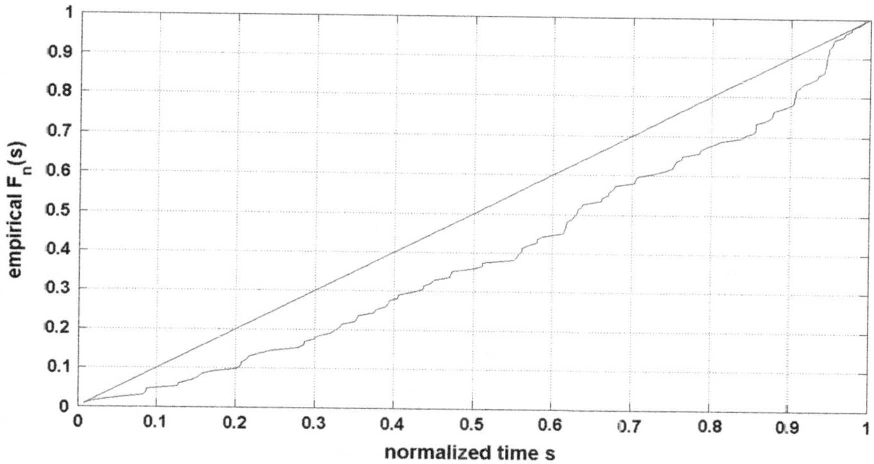

Fig. 3.24 The injured by earthquakes. Catalog of earthquake victims (Japan), 1900–2012. The empirical distribution function $F_n(t)$ of normalized occurrence times τ_j compared with the uniform distribution function (*diagonal line*)

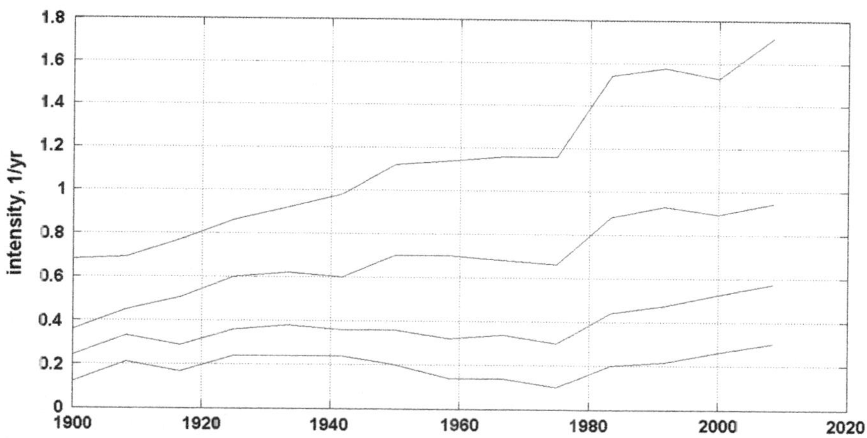

Fig. 3.25 The injured by earthquakes. Catalog of earthquake victims (Japan), 1900–2012. Intensities for four thresholds from bottom to top: $h_1 = 1$ ($n_1 = 131$); $h_2 = 10$ ($n_2 = 78$); $h_3 = 100$ ($n_3 = 44$); $h_4 = 400$ ($n_4 = 23$), smoothed by 15-year time window

The Kolmogorov distance KD = 1.21 which corresponds to p–v = 0.11. This value is on the border of acceptance, but still more than 0.10 and we can accept with some reservation the hypothesis of the stationarity of the event times for $h = 400$. Now we check for stationarity the distribution of injured. Figure 3.27 shows the cumulative sums of $\log(x_j)$, x_j—numbers of injured. We see that

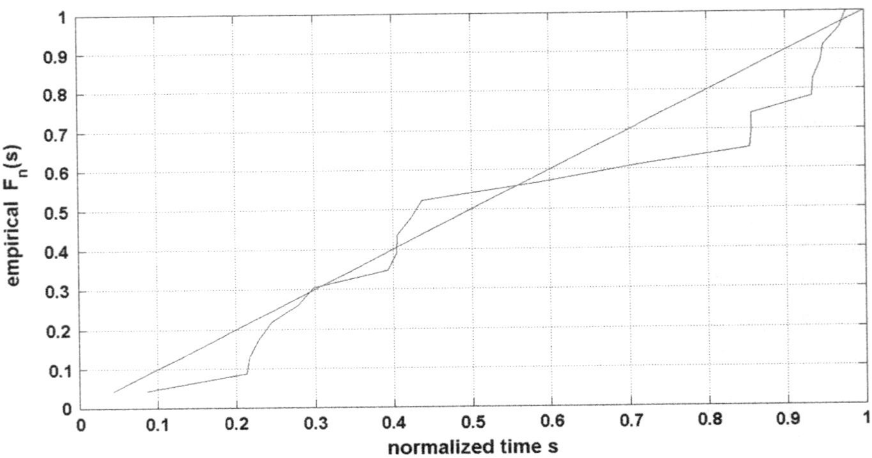

Fig. 3.26 The injured by earthquakes. Catalog of earthquake victims (Japan), 1900–2012. The empirical DF of normalized event times for $h = 400$. The empirical distribution function F_n(s) of normalized occurrence times s_j compared with the uniform distribution function (*diagonal line*)

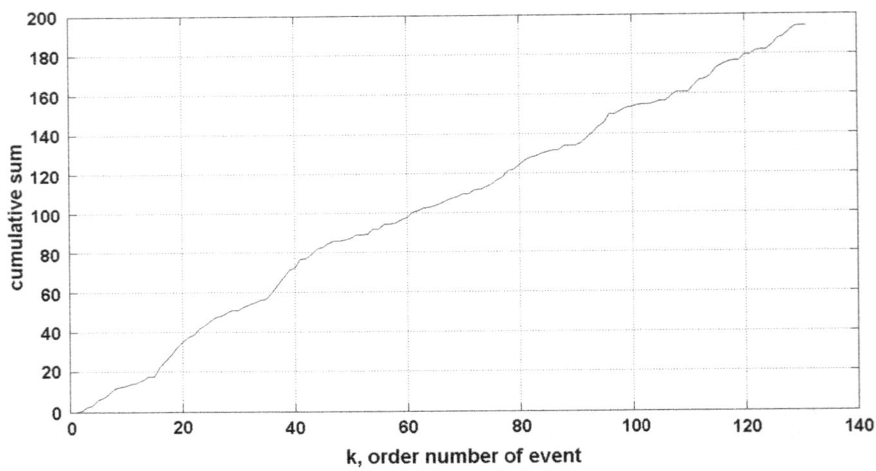

Fig. 3.27 The injured by earthquakes. Catalog of earthquake victims (Japan), 1900–2012. The cumulative sums of $\log(x_k)$, x_k—numbers of injured

deviations from a straight line are not significant, and we can accept the hypothesis of the stationarity of distribution of injured. So, we accept the hypothesis of the stationarity of this catalog.

The tail graph of injured is shown on Fig. 3.28. We see that on the whole the tail 1− F(x) has power-like behavior with possible increasing inclination. So, it is reasonable to take logarithms and to analyze log(x). Figure 3.30 shows the

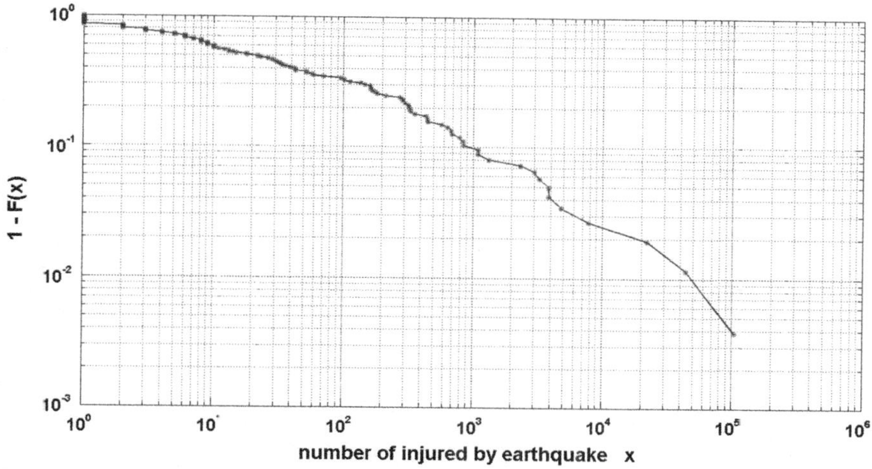

Fig. 3.28 The injured by earthquakes. Catalog of earthquake victims (Japan), 1900–2012. The tail graph $1-F(x)$

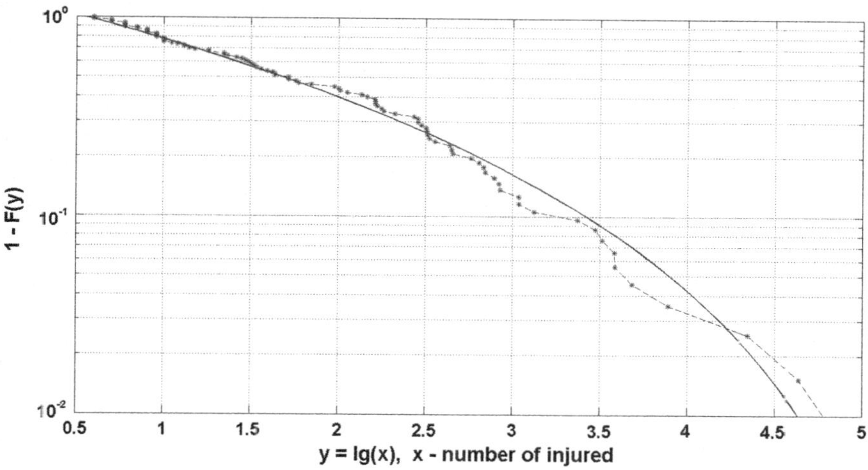

Fig. 3.29 The injured by earthquakes. Catalog of earthquake victims (Japan), 1900–2012. The extreme tail $1-F(\mathrm{y})$ ($y = \log(x)$, x—number of injured by earthquake) and approximating GPD-tail: $h = 0.55$; $\xi = -0.374$; $s = 1.860$; $n = 99$

contribution of the p-fraction of the largest events to the total sum. We see that 10 % of the most disastrous events are responsible for 94 % of the total number of injured. Such sample can be characterized as sample with a strong concentration. Figure 3.29 shows the extreme part of the tail used for parameter estimation along with fitted GPD-curve. We see that the GPD-approximation is on the whole satisfactory in spite of some oscillations.

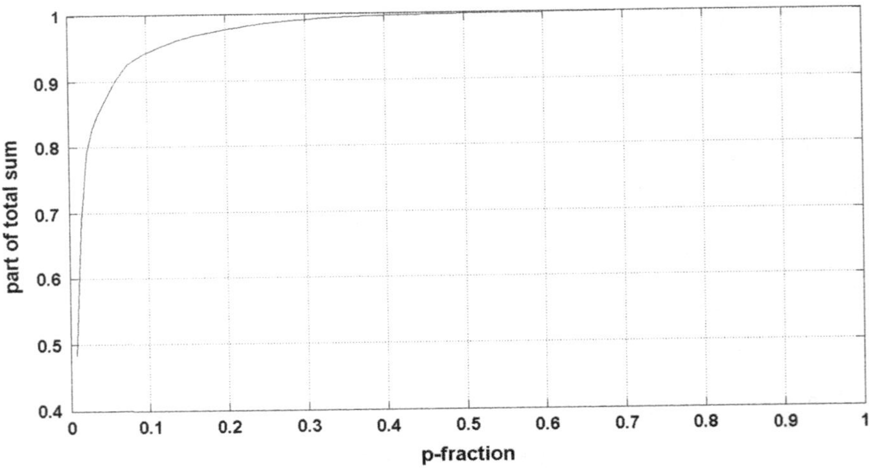

Fig. 3.30 The injured by earthquakes. Catalog of earthquake victims (Japan), 1900–2012. The contribution of the p-fraction of the most deadly events to the total death toll

The Kolmogorov test for a grid of thresholds is shown on Fig. 3.31. The best fit of GPD-distribution under $h = 0.55$ $(\xi = -0.374 \pm 0.063;\ s = 1.86 \pm 0.30)$ provides $p\text{–}v = 0.69$, which supports GPD-distribution.

Now we are able to calculate the quantiles $Q_q(\tau)$. Figure 3.32 shows the GPD-quantiles for 3 different confidence levels $q = 0.90;\ 0.95;\ 0.99$. Again we see jumps for very small τ corresponding to absence of events on time interval $[0;\ \tau.]$

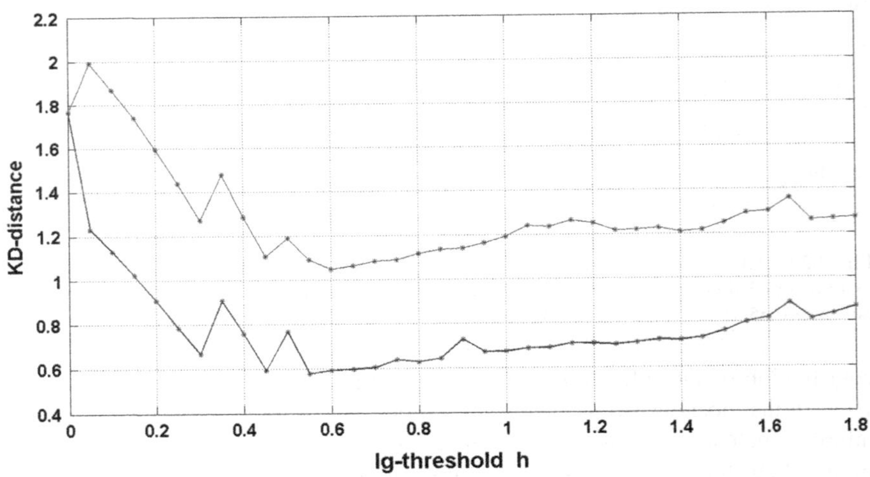

Fig. 3.31 The injured by earthquakes. Catalog of earthquake victims (Japan), 1900–2012. The KD-distances for a grid of log-thresholds h_j. *Thick line*—GPD-fitting; *thin line*—ED-fitting

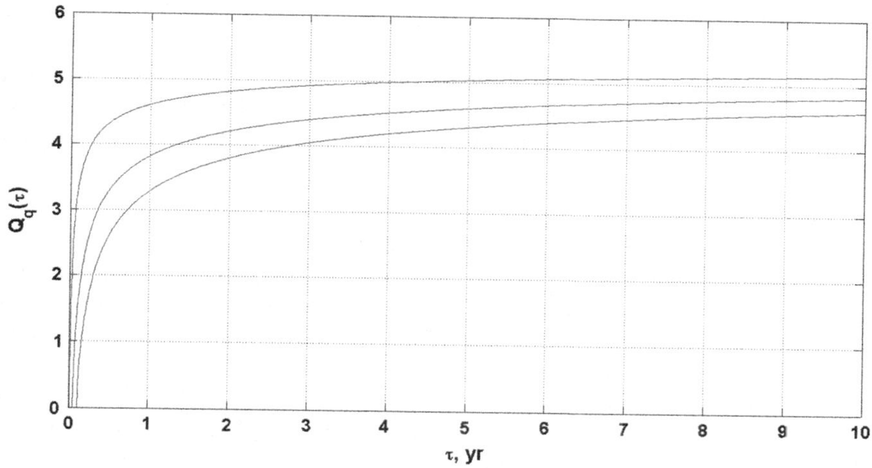

Fig. 3.32 The injured by earthquakes. Catalog of earthquake victims (Japan), 1900–2012. The GPD-quantiles for three different confidence levels: $q = 0.90$ (*lower curve*); 0.95 (*middle curve*); 0.99 (*upper curve*)

The "maximum possible size" (rightmost limit M_{max} of the GPD) is:

$$M_{max} = h - s/\xi = 5.52 \text{ (this corresponds to 330,000 injured persons)}$$

This is much more than the quantile $Q_{0.95}(10) = 4.81$ (65,000).

3.3 Floods (Victims, Overall Economic Losses)

3.3.1 Cautions on the Accuracy of the Flood Damage Data

Flood damage estimates are reported in many different ways, and are subject to a wide variety of errors. Estimates come from federal, state, or county level government officials. Some inaccuracies and mistakes in the data are inevitable, damages are often underreported. Besides, different definitions of term "flood" are used. One of the most critical discrepancies of these data occurs with storm surge related flooding caused by tropical cyclones. Coastal flooding caused by storm surge is not counted in the figures of the flood damage data used below. The record season of 2005, with hurricanes Katrina and Rita, were undoubtedly enormous flooding events. However, the damages associated with hurricane Katrina were largely due to storm surge, and not fresh water flooding (associated to rainfall). Therefore, the annual figure of \$51B for water year 2005, although much higher than any other year, does not account for most of the flooding produced by Katrina. On the other hand, the damages from hurricanes that we shall analyze below, include the hurricane Katrina with \$108B damage, which contributed in \$141B of annual US hurricane damage, 2005.

In this section we shall use the data of the International Disaster Database (www.emdat.be/). The data consist of three types of flood losses in USA, 1900-2011: fatalities, numbers of affected by flood and estimated economic losses from flood.

(a) ***Fatalities***

The catalog contains fatality data of 99 floods. Figure 3.33 shows general view of these data. Looking at Fig. 3.33 we see that the intensity of events sharply increases since 1995. Perhaps, this increase can be explained by more careful registration of victims in later times. On Fig. 3.34 the intensity of events is shown for 1995–2011. We see a stable behavior of the event flow. So, we shall use below data from 1995 onwords.

On Fig. 3.35 we compare the empirical DF of normalized time moments s_i. $\hat{F}_n(s)$ with uniform DF. The deviations from diagonal are small. The Kolmogorov distance $KD = 0.633$ corresponds to p–value $= 0.82$. Since this p-value (p–v) is considerably more than 0.1, we can accept the hypothesis of stationarity of t_i.

Now we consider the fatalities figures. The tail graph is shown on Fig. 3.36. The tail decrease goes rather gradually at the middle range (which is typical of the Pareto distribution), but accelerates at $x > 15$, approaching the exponential tail behavior. Thus, it is not clear, whether the log-transformation of data is appropriate? KD-distances, characterizing goodness of fit are shown on Fig. 3.38 for a grid of thresholds. We see that the original data have been fitted by ED much better ($h = 10^{0.45} = 2.5$; $n = 41$; $KD = 0.836$; p–$v = 0.22$) than logarithms ($h = 10^{0.7} = 5$; $n = 32$; $KD = 1.515$; p–$v = 0.0004$). Thus, we use the original data for further processing. Figure 3.39 shows the contribution of the p-fraction of

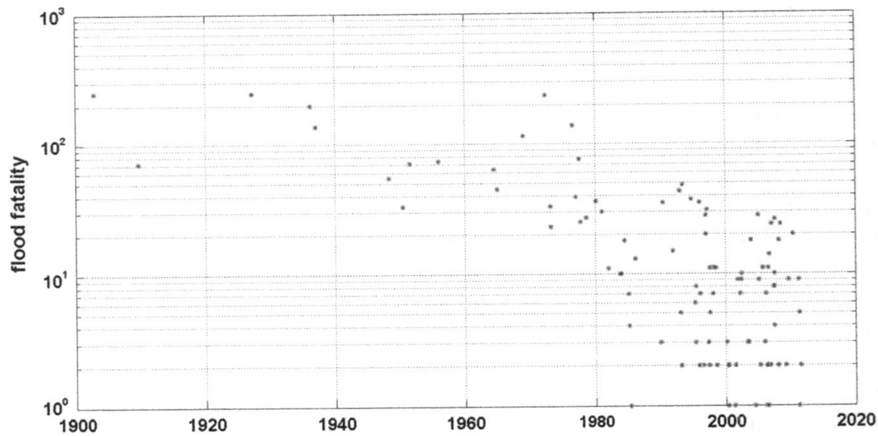

Fig. 3.33 The fatality events of the catalog of flood victims, USA, 1900–2011

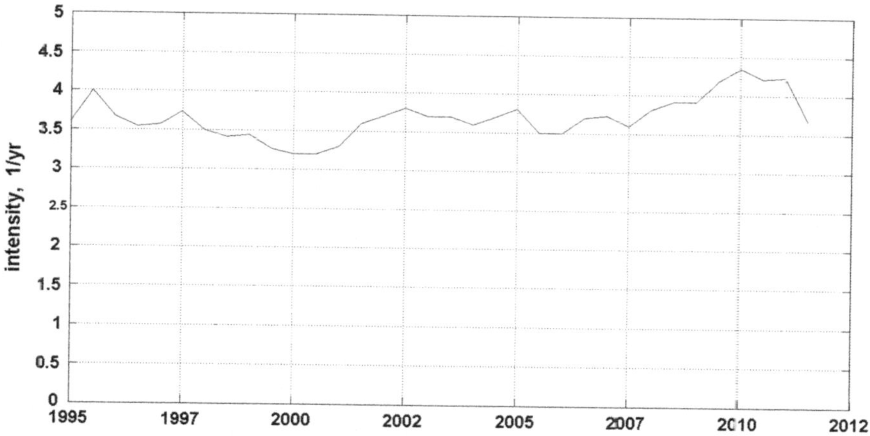

Fig. 3.34 The intensity of floods with fatalities, USA, 1995–2011, smoothed by 5-year time window

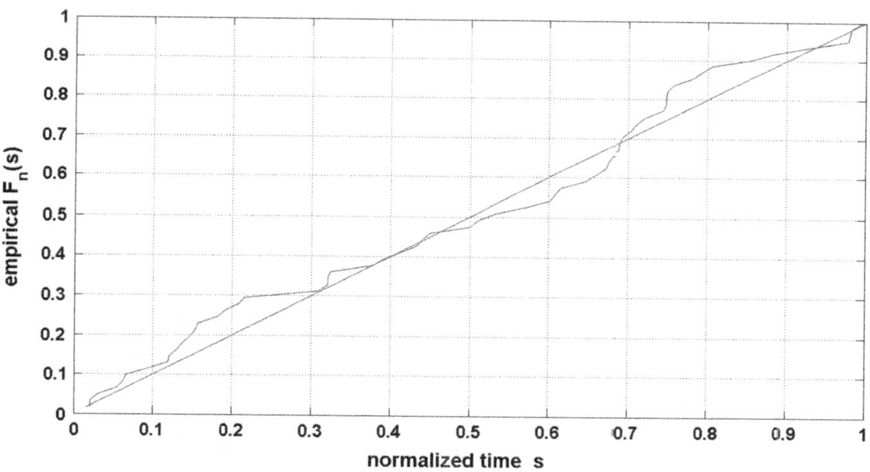

Fig. 3.35 Flood fatalities, USA, 1995–2011. The empirical DF of normalized event times s_i $\hat{F}_n(s)$

the most deadly events to the whole death toll. We see that 10 % of the most deadly events are responsible for 55 % of the death toll. Such a sample can be characterized as a sample with a weak concentration. Figure 3.37 shows the extreme part of the tail used for parameter estimation along with fitted exponential curve. We see that behavior of the extreme part of sample tail is rather unstable, and an exponential curve gives the best possible approximation in this complicated situation. The p-value 0.22 is not too high, but still it gives the ground to accept the ED.

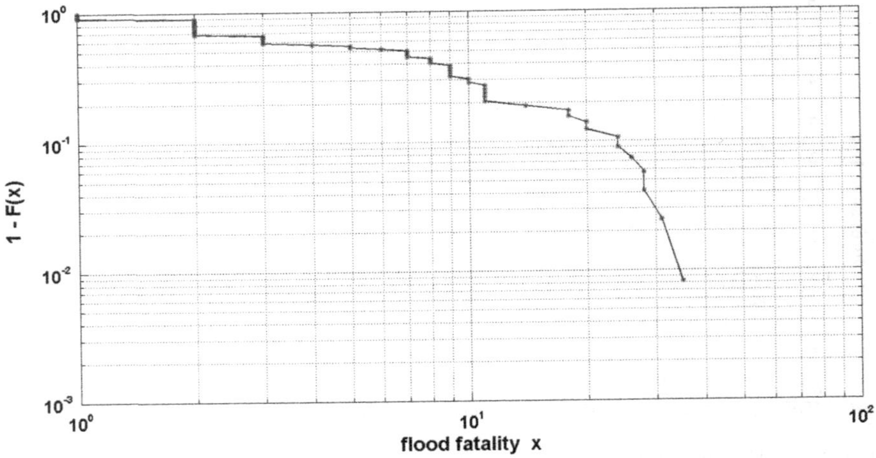

Fig. 3.36 Flood fatalities, USA, 1995–2011. The tail graph 1–$F(x)$

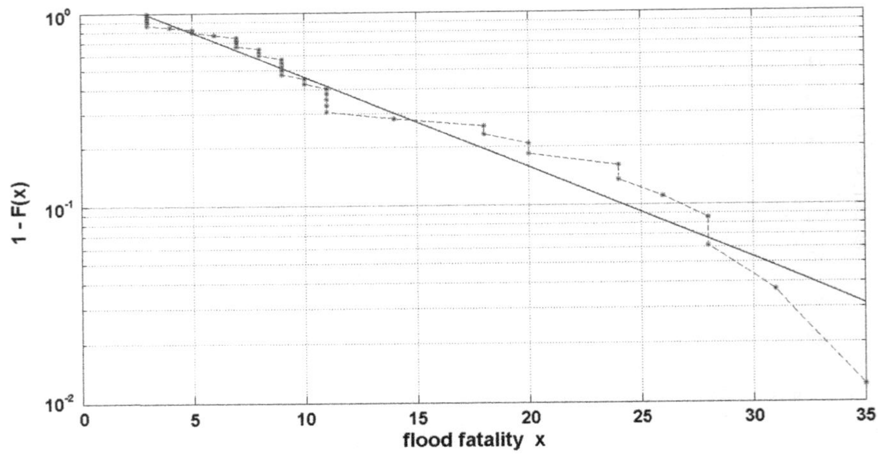

Fig. 3.37 Flood fatalities, USA, 1995–2011. The extreme tail 1–$F(x)$, and approximating EXP-tail $\exp[-\alpha\cdot(x-h)]$; $h = 0.45$; $\alpha = 0.108$; $n = 41$

On Fig. 3.40 the cumulative sums of fatalities for 1995–2011 ($n = 61$) are shown. In spite of some local fluctuations the trend on the whole looks like a linear function. If there are some deviations they seem to be insignificant. So, we accept the hypothesis of the stationarity of this catalog. Now we are able to calculate the quantiles $Q_q(\tau)$, using ED fitting. Figure 3.41 shows the ED-quantiles for three different confidence levels $q = 0.90$; 0.95; 0.99.

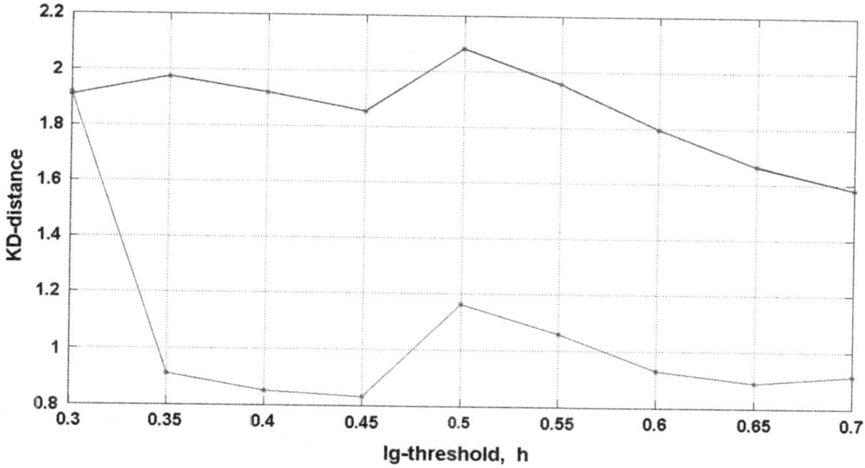

Fig. 3.38 Flood fatalities, USA, 1995–2011. The KD-distances for a grid of log-thresholds h_j. *Thick line*—GPD-fitting; *thin line*—ED-fitting

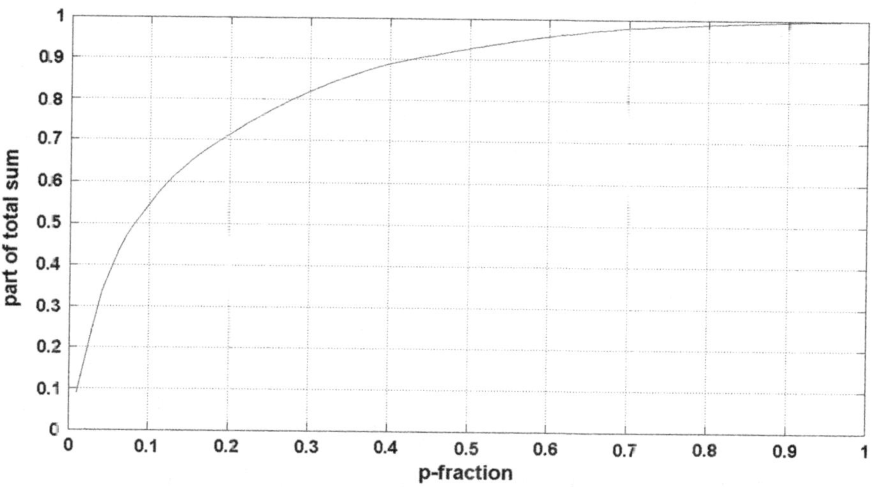

Fig. 3.39 Flood fatalities, USA, 1995–2011. The contribution of the *p*-fraction of the most deadly events to the whole death toll

(b) *Affected by floods*

Affected people, as defined in EM-DAT, are people who require immediate assistance during a period of emergency, including displaced or evacuated people. The catalog contains numbers of affected in 95 floods, 1970–2011. Figure 3.42 shows these numbers in time. The maximum number (11,000,148) occurred at

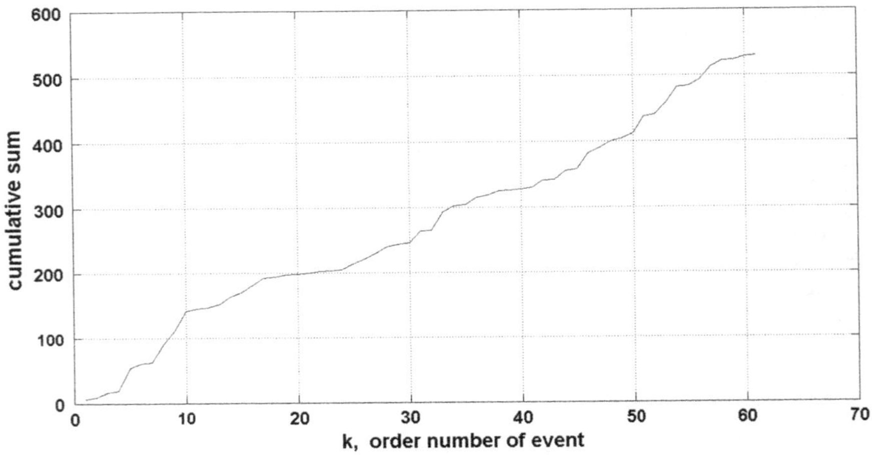

Fig. 3.40 Flood fatalities, USA, 1995–2011. The cumulative sums of fatalities for 1995–2011 (n = 61) are shown

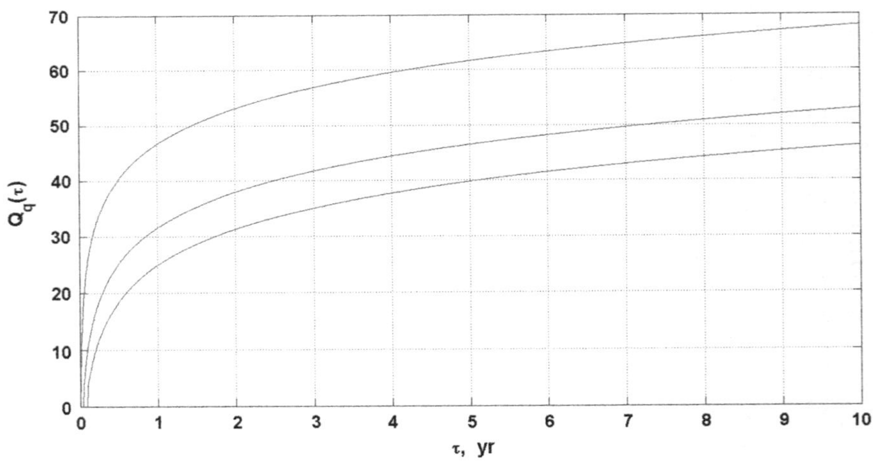

Fig. 3.41 Flood fatalities, USA, 1995–2011. ED-quantiles for three different confidence levels: $q = 0.90$ (*lower curve*); 0.95 (*middle curve*); 0.99 (*upper curve*)

09.06.2008. Looking at Fig. 3.42 we see that the intensity of events sharply increases at 1995, which coincides with mentioned above behavior of flood fatalities. On Fig. 3.43 the intensity of events is shown for 1995–2011. We see a stable behavior of the event flow. So, we have used the data from 1995 onwards.

On Fig. 3.44 we compare the empirical DF of normalized time moments s_i. $\hat{F}_n(s)$ with uniform DF. The deviations from diagonal are relatively small. The

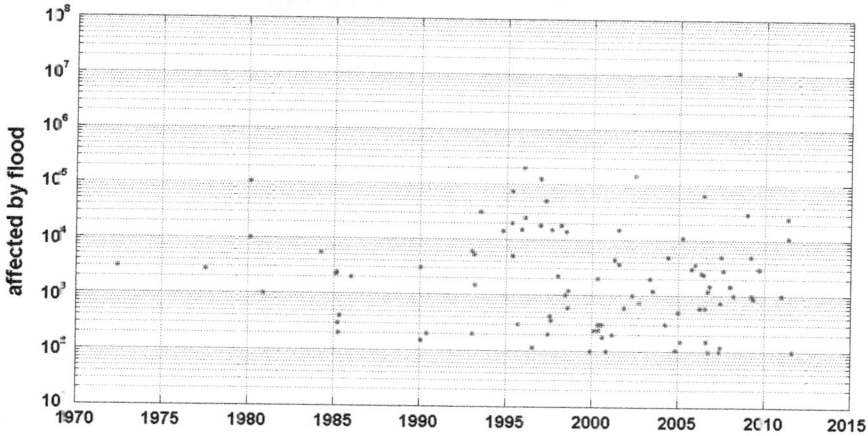

Fig. 3.42 Affected in floods, USA, 1995–2011. Time–event size diagram

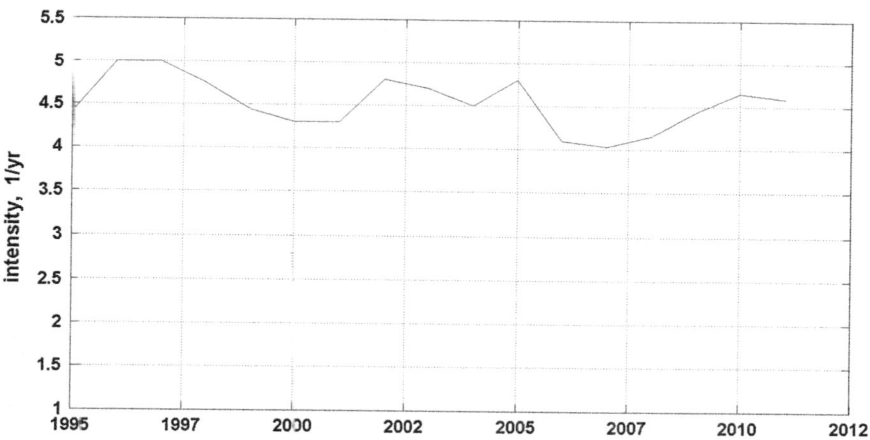

Fig. 3.43 Affected in floods, USA, 1995–2011. The intensity of events for 1995–2011, smoothed by 10-year time window

Kolmogorov distance $D_n = 0.692$ corresponds to p-value $= 0.72$. Since this p-value (p–v) is more than 0.1, we can accept the hypothesis of stationarity of t_i.

Now we consider the figures of affected by floods. The tail graph is shown on Fig. 3.45. We see that on the whole the tail 1 – F(x) has power-like behavior with one outlier (maximum event May 09, 2008 with 11,000,148 affected). So, it is reasonable to take logarithms and to analyze log(x). Figure 3.47 shows the contribution of the p-fraction of the largest events to the total sum. We see that 10 % of the most disastrous events are responsible for 97.7 % of the total number of affected. This is a tail with "strong concentration".

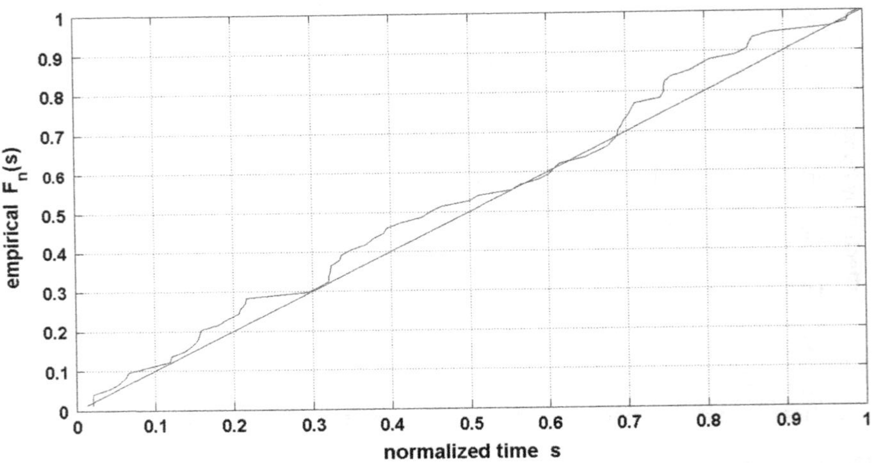

Fig. 3.44 Affected in floods, USA, 1995–2011. The empirical DF $\hat{F}_n(s)$ of normalized event times s_i

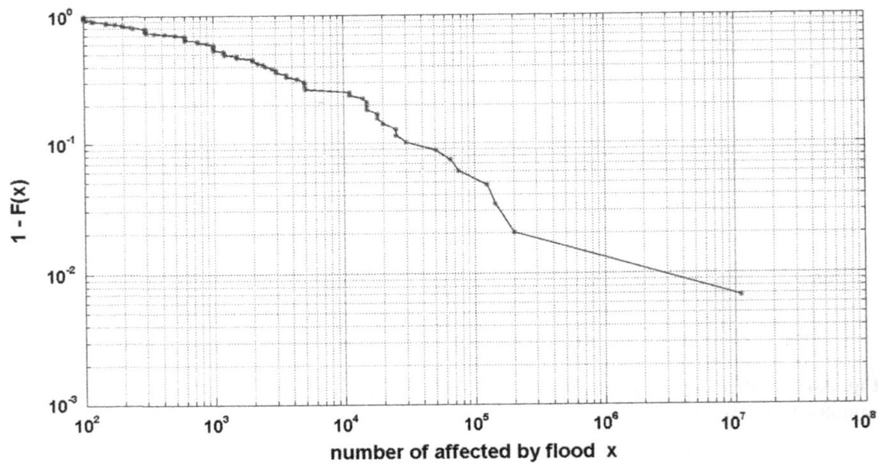

Fig. 3.45 Affected in floods, USA, 1995–2011. The tail graph 1 – (Fx)

The Kolmogorov test for a grid of thresholds is shown on Fig. 3.48. We see that minimum KD-distance 0.567 is reached at $h = 2.7$ by GPD. Using the simulation method we got $p\text{–}v = 0.665$, which allows us to accept GPD-distribution ($h = 2.7$; $\xi = -0.182 \pm 0.113$; $s = 1.205 \pm 0.302$). Figure 3.46 shows the extreme part of the tail used for parameter estimation along with fitted GPD-curve. We see that the approximation is more or less satisfactory.

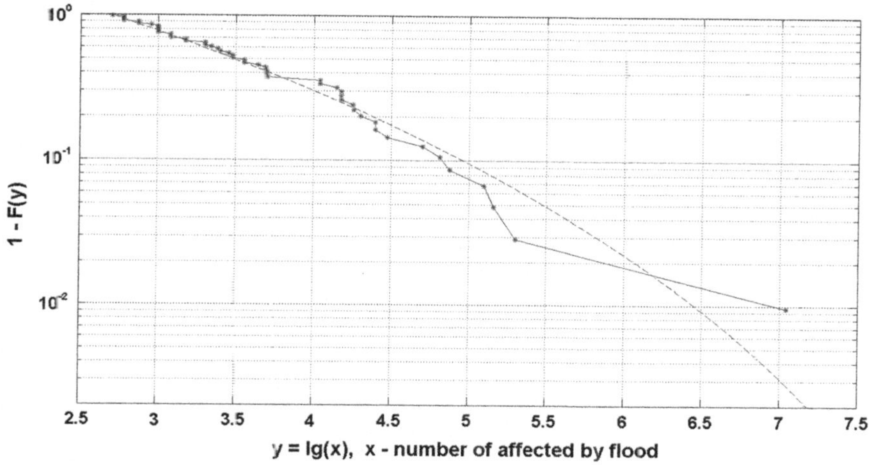

Fig. 3.46 Affected in floods, USA, 1995–2011. The extreme tail $1 - F(y)$ ($y = \log(x)$, x—number of affected in flood) and approximating GPD-tail: $h = 2.7$; $\xi = -0.182$; $s = 1.205$; $n = 52$

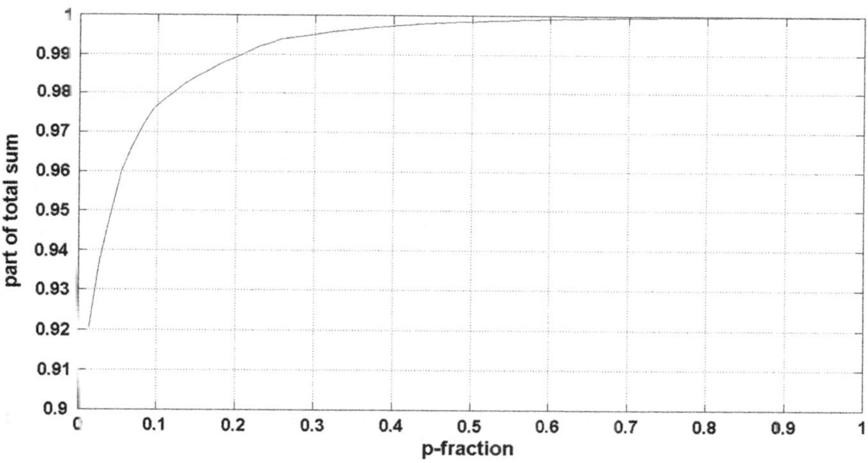

Fig. 3.47 Affected in floods USA, 1995–2011. The contribution of the p-fraction of the most disastrous events to the total sum of affected

Now we are able to calculate the quantiles $Q_q(\tau)$. We use for these quantiles GPD-distribution as providing the best fitting. Figure 3.49 shows GPD-quantiles for three different confidence levels $q = 0.90; 0.95; 0.99$.

The "maximum possible size" (rightmost limit M_{max} of the GPD) is:

$$M_{max} = h - s/\xi = 9.31(2 \cdot 10^9 \text{peoples}).$$

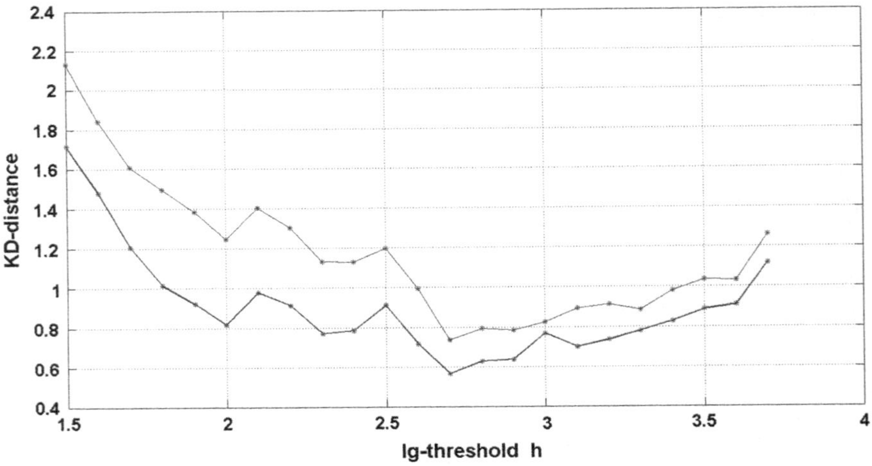

Fig. 3.48 Affected in floods USA, 1995–2011. The KD for a grid of log-thresholds. *Thick line*—GPD-fitting; *thin line*—ED-fitting

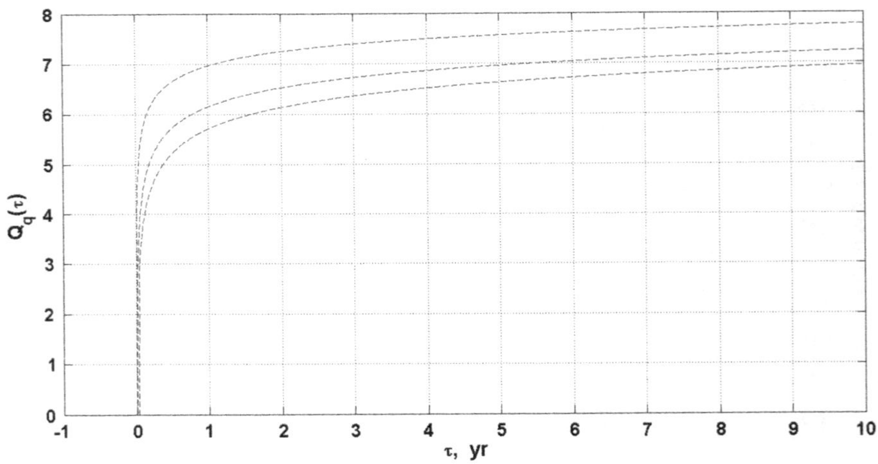

Fig. 3.49 Affected in floods USA, 1995–2011. GPD-quantiles for three different confidence levels: $q = 0.90$ (*lower curve*); 0.95 (*middle curve*); 0.99 (*upper curve*)

Again, we can say that such gigantic figure hardly has a practical value. In the same time the quantile $Q_{0.95}(10) = 7.25$ ($17.7 \cdot 10^6$ peoples) looks quite realistic.

(c) *Estimated damages caused by floods in USA*

The catalog contains estimated economic loss data of 78 floods in USA, 1900–2011 (indexed to 2011) in millions of USA \$. Figure 3.50 shows general

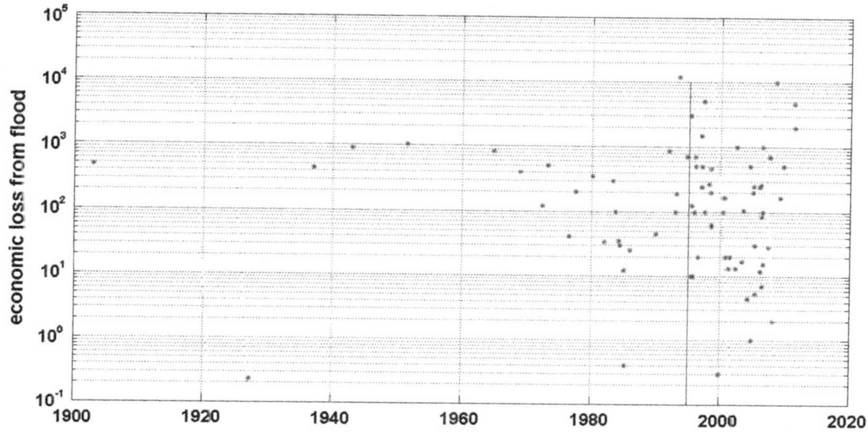

Fig. 3.50 Estimated economic losses from floods, USA, 1995–2011, in 10^6 \$ (adjusted to 2011). Time–event size diagram

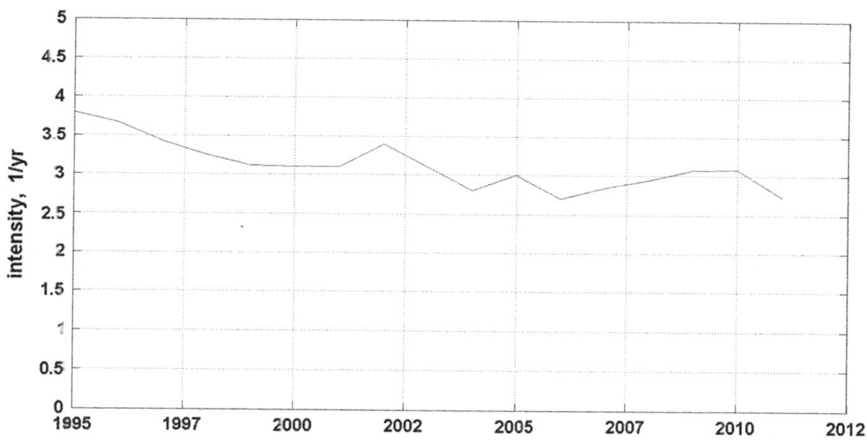

Fig. 3.51 Estimated economic losses from floods, USA, 1995–2011, in 10^6 \$. The intensity of events, smoothed by 10-year time window

view of these data. Looking at Fig. 3.50 we see that the intensity of events sharply increases since 1995 (shown by vertical line on the Figure), which is consistent with behavior of data on fatalities and affected. Again, we shall analyze data only from 1995 onwards.

On Fig. 3.51 the intensity of events is shown for 1995–2011. We see a stable behavior of the event flow with a weak decrease to the end of interval.

On Fig. 3.52 we compare the empirical DF of normalized time moments s_i. $\hat{F}_n(s)$ with uniform DF. The Kolmogorov distance $D_n = 1.062$ corresponds to

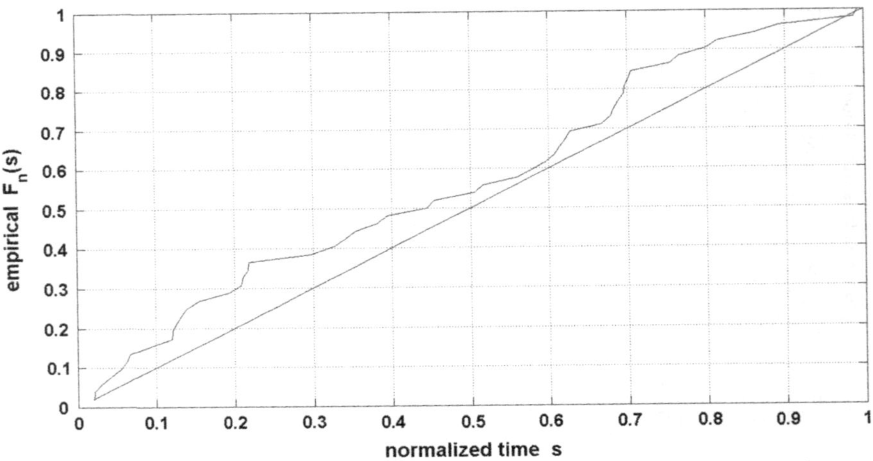

Fig. 3.52 Estimated economic losses from floods, USA, 1995–2011, in 10^6 \$. The empirical DF $\hat{F}_n(s)$ of normalized time moments s

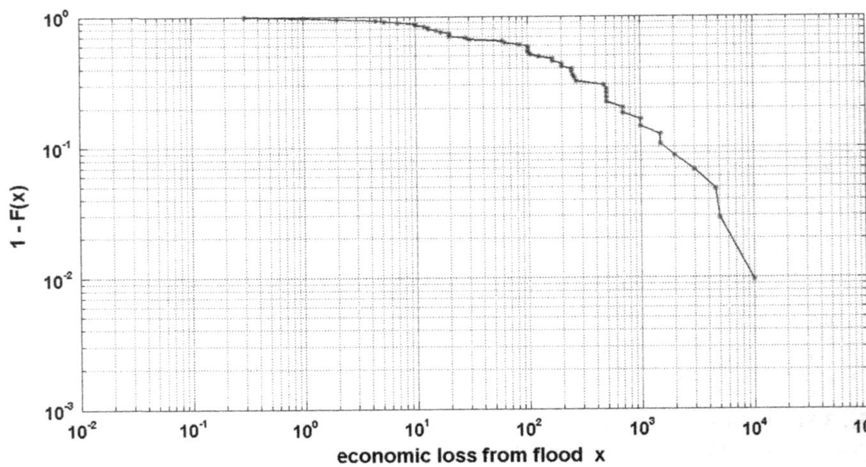

Fig. 3.53 Estimated economic losses from floods, USA, 1995–2011, in 10^6 \$. The tail graph $1 - F(x)$

p-value $= 0.21$. Since this p-value is more than 0.1, we can accept the hypothesis of stationarity of t_i.

Now we consider the loss figures. The tail graph is shown on Fig. 3.53. The tail decrease goes rather gradually which is typical for the Pareto distribution. So, we take logarithms for further analysis. KD-distances, characterizing goodness of fit

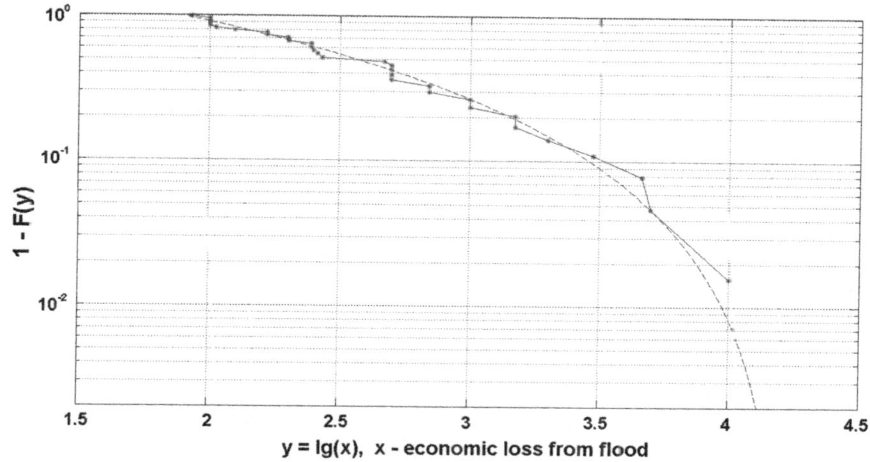

Fig. 3.54 Estimated economic losses from floods, USA, 1995–2011, in 10^6 \$. The extreme tail $1 - F(y)$ ($y = \log(x)$, x—loss in 10^6 USD) and approximating GPD-tail: $h = 1.9$; $\xi = -0.486$; $s = 1.129$; $n = 32$

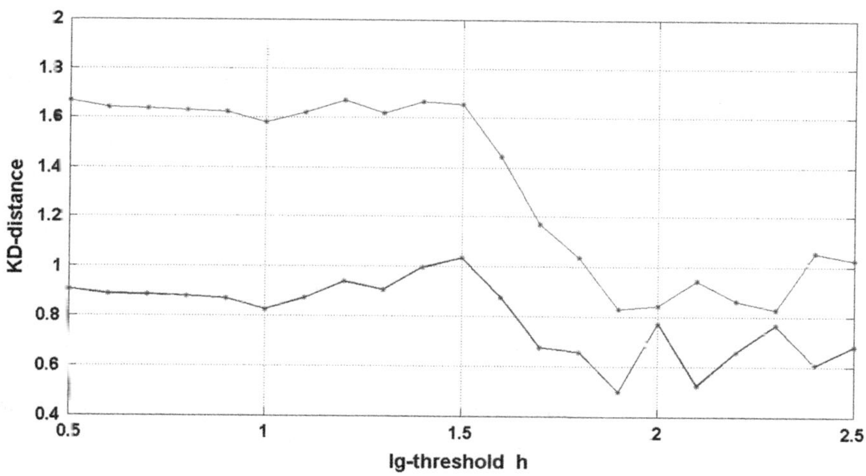

Fig. 3.55 Estimated economic losses from floods, USA, 1995–2011, in 10^6 \$. KD-distances for a grid of log-thresholds. *Thick line*—GPD-fitting; *thin line*—ED-fitting

are shown on Fig. 3.55 for a grid of thresholds. The threshold $h = 1.9$ provides the best fitting for GPD ($KD = 0.501$; $\xi = -0.486 \pm 0.091$; $s = 1.129 \pm 0.286$).

Figure 3.56, shows the contribution of the p-fraction of the most costly events to the whole sum of losses. We see that 10 % of the most deadly events are responsible for 68 % of the total loss. Such sample can be characterized as sample

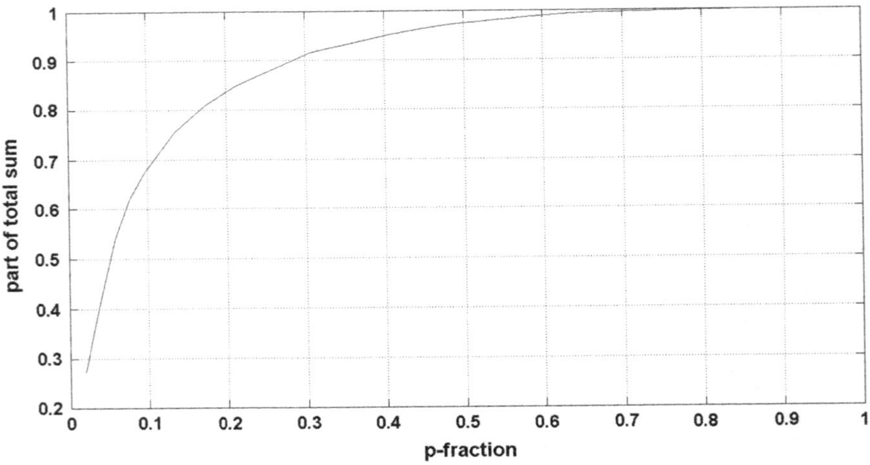

Fig. 3.56 Estimated economic losses from floods, USA, 1995–2011, in 10^6 \$. The contribution of the *p*-fraction of the most costly events to the total sum of all losses

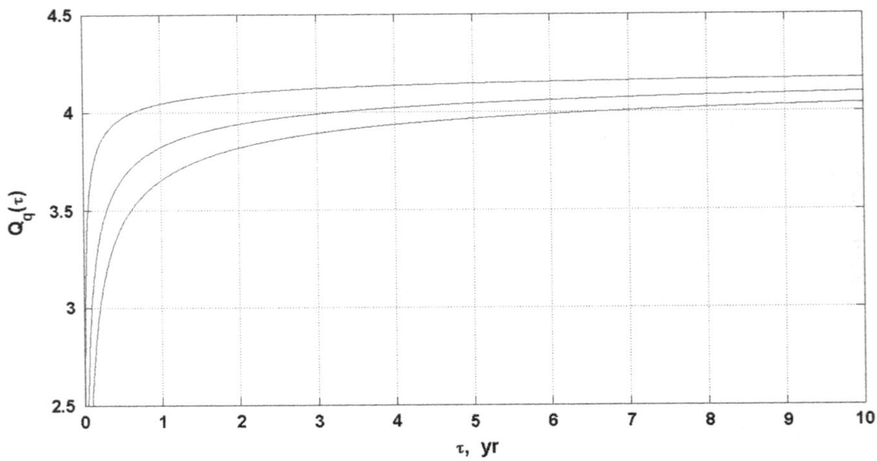

Fig. 3.57 Estimated economic losses from floods, USA, 1995–2011, in 10^6 \$. GPD-quantiles for three different confidence levels: $q = 0.90$ (*lower curve*); 0.95 (*middle curve*); 0.99 (*upper curve*)

with a "moderate concentration". Figure 3.54 shows the extreme part of the tail used for parameter estimation along with fitted GPD-curve. We see that the approximation is quite satisfactory.

Now we calculate the quantiles $Q_q(\tau)$, using GPD fitting. Figure 3.57 shows the GPD-quantiles for three different confidence levels $q = 0.90$; 0.95; 0.99.

3.4 Tornadoes (Fatalities)

In this section we shall use the data of tornado fatalities displayed in the Internet (http://en.wikipedia.org/wiki/List_of_North_American_tornadoes_and_tornado_outbreaks#1900.E2.80.931919). The data consist of fatality numbers of 249 tornadoes, USA, 1900–2012. Figure 3.58 shows general view of these data. Looking at Fig. 3.58 we see that the intensity of events depends on their values. On Fig. 3.59 the intensity of events is shown for two thresholds: $h_1 = 1$ ($n_1 = 249$) and $h_2 = 10$ ($n_2 = 143$). We see that the threshold h_2 provides event flow with rather stable intensity whereas there are $n = 143$ events exceeding this threshold.

On Fig. 3.60 we compare the empirical distribution functions $F_{n_1}^{(1)}(x)$, $F_{n_2}^{(2)}(x)$ of normalized time moments corresponding to these two thresholds with uniform DF. The deviations of $F_{n_1}^{(1)}(x)$ from diagonal are large (corresponding $KD = 3.69$; p–$v = 3 \cdot 10^{-12}$), whereas the deviations of $F_{n_2}^{(2)}(x)$ are much smaller (corresponding $KD = 1.08$; p–$v = 0.20$). Since the last p–v is more than 0.1 we can accept the hypothesis of stationarity of time moments t_i for events exceeding $h_2 = 10$.

Now we check stationarity of the fatalities x_j exceeding $h_2 = 10$. Figure 3.61 shows the cumulative sums of $\log_{10}(x_j)$. Somewhere near $t_j = 60$ (corresponding to May 09, 1953) we can distinguish a small but clear decrease of slope. In order to clarify the situation we plotted two sample DF: $G_{m_1}^{(1)}(x)$ and $G_{m2}^{(2)}(x)$, relating to $t \leq$ May 09, 1953, ($m_1 = 60$) and to $t >$ May 09, 1953 ($m_2 = 83$) correspondingly, see Fig. 3.62. We see that distribution functions differ quite definitely, in particular in the middle range. The Kolmogorov–Smirnov distance (KSD)

$$KSD = \sqrt{\frac{m_1 m_2}{m_1 + m_2}} \max \left| G_{m_1}^{(1)}(x) - G_{m_2}^{(2)}(x) \right|$$

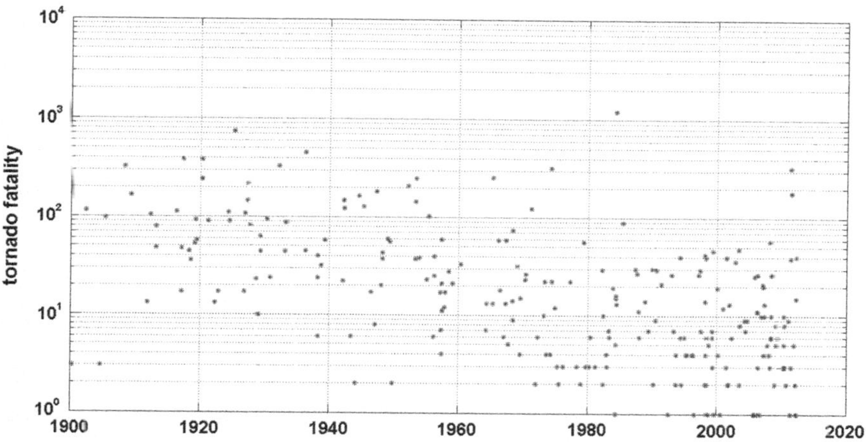

Fig. 3.58 Tornado fatalities, USA, 1900–2012. Time–event size diagram

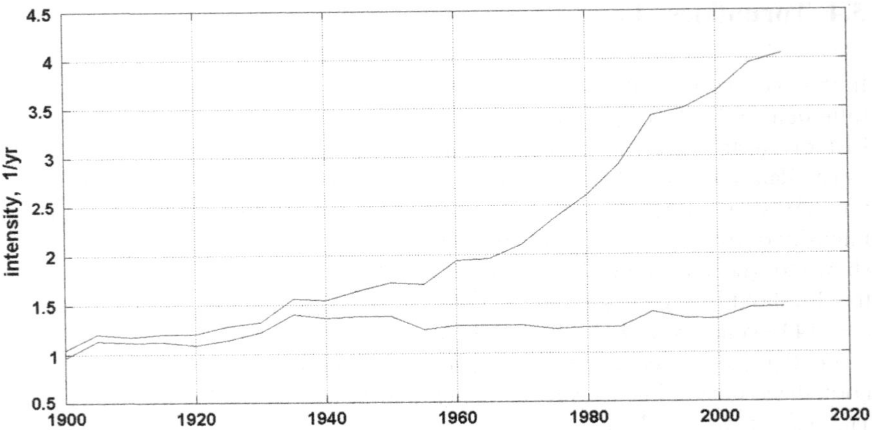

Fig. 3.59 Tornado fatalities, USA, 1900–2012. The intensities smoothed by 40-year time window for two lower thresholds: $h_l = 1$ ($n = 249$), *upper curve;* $h_2 = 10$ ($n = 143$), *lower curve*

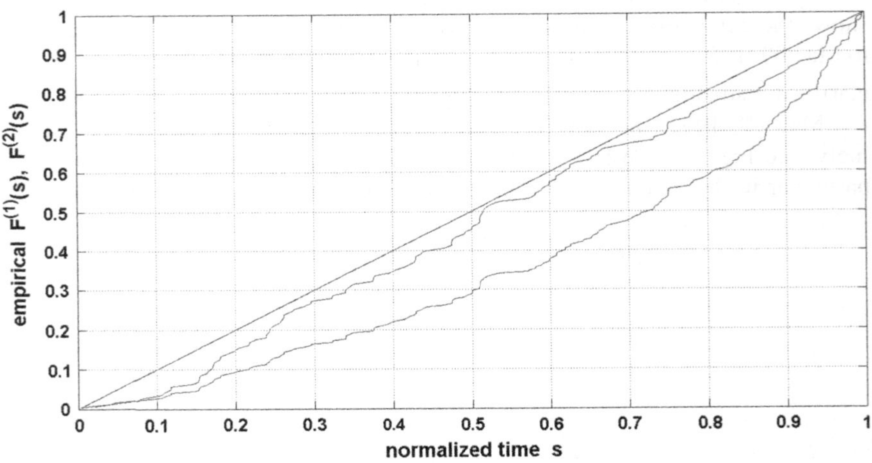

Fig. 3.60 Tornado fatalities, USA, 1900–2012. Two empirical distribution functions $F_{n_1}^{(1)}(x)$, $F_{n_2}^{(2)}(x)$ of normalized time moments corresponding to thresholds $h_l = 1$ ($n = 249$), *lower curve;* $h_2 = 10$ ($n = 143$), *upper curve*

equals to 3.02 which corresponds to p–$v = 1 \cdot 10^{-8}$. So, we have to use for the further analysis only 83 data observed after May 09, 1953.

The tail graph is shown on Fig. 3.63. The tail is typical of the Pareto distribution. So, we use logarithmically transformed data. KD-distances, characterizing goodness of fit are shown on Fig. 3.65 for a grid of thresholds. We see that the original data

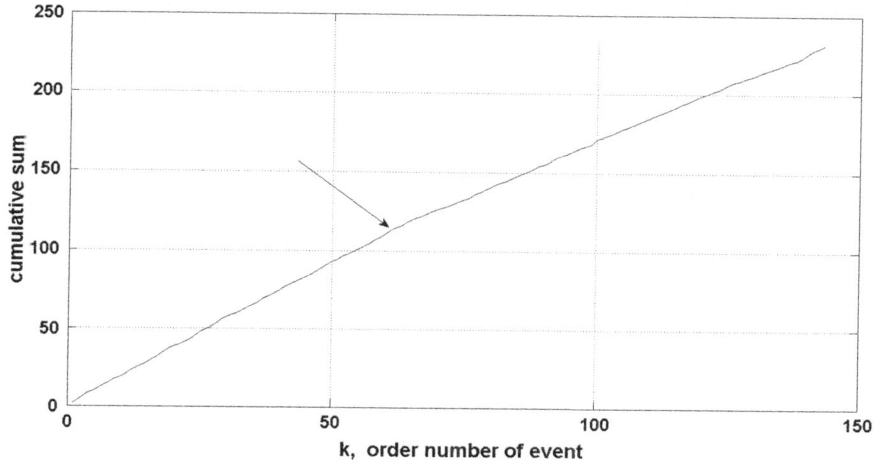

Fig. 3.61 Tornado fatalities, USA, 1900–2012. The cumulative sums of $\log_{10}(x_k)$, x_k—number of dead. The arrow indicates slope break (at time 09.05.1953)

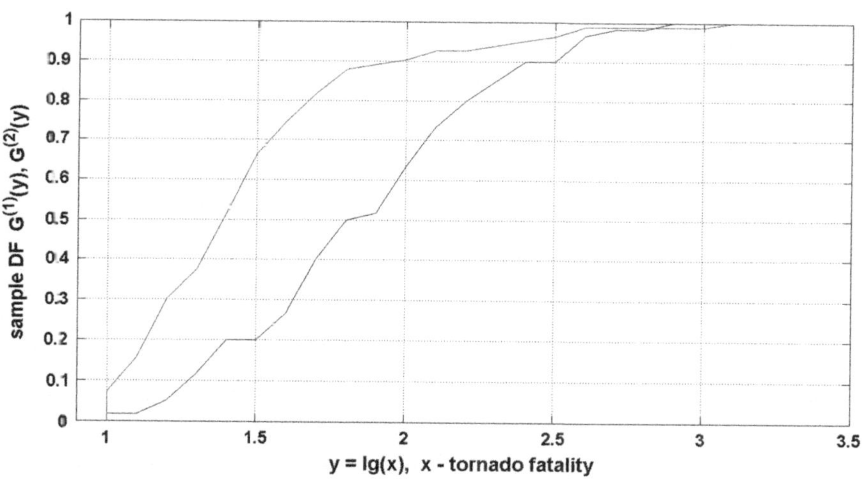

Fig. 3.62 Tornado fatalities, USA, 1900–2012. Two sample DF: $G^{(1)}_{m_1}(x)$ and $G^{(2)}_{m2}(x)$, relating to $t \leq$ May 09, 1953, ($m_1 = 60$), *right curve*, and to $t >$ May 09, 1953 ($m_2 = 83$), *left curve*

have been fitted by both distributions (ED and GPD) almost identically for thresholds $h = 1.3 - 1.35$. The form parameter of GPD for these thresholds was found to be practically zero ($\sim 10^{-8}$). Thus, GPD and ED in this case are practically identical. The best fit corresponds to the threshold $h = 1.3$; $n = 53$. The parameter of exponential distribution $\alpha = 2.75 \pm 0.38$. As we remarked above, the exponential distribution for $\log(x)$ means the Pareto distribution for original x with parameter

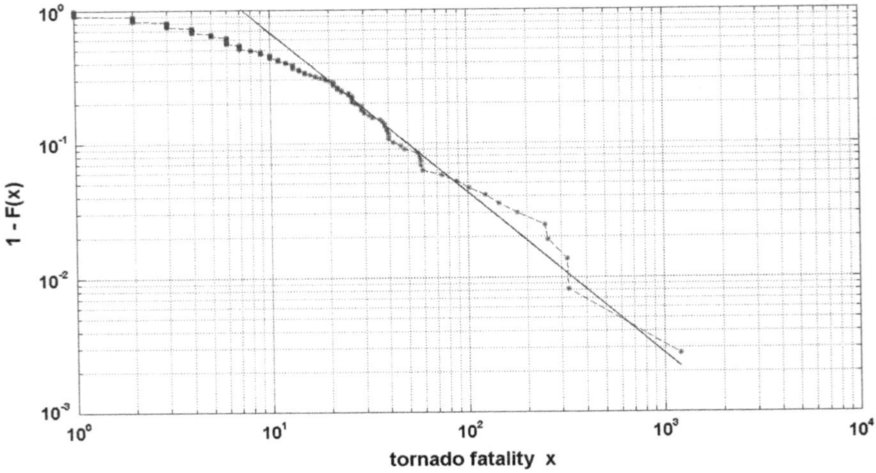

Fig. 3.63 Tornado fatalities, USA, 1953–2012, threshold $h \geq 10$, $n = 83$. The tail function $1 - F(x)$. Approximating line is the same as on Fig. 3.64

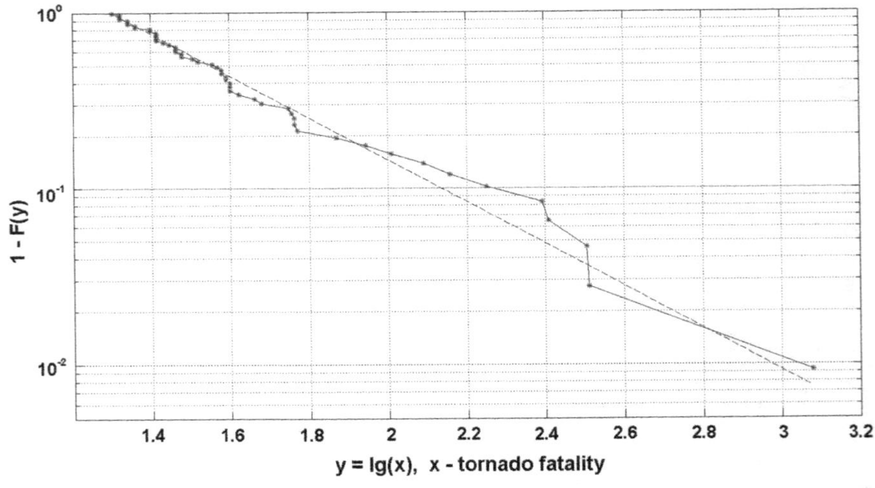

Fig. 3.64 Tornado fatalities, USA, 1953–2012. The extreme tail $1 - F(y)$ ($y = \log(x)$, x—number of dead) and approximating EXP-tail $exp[-\alpha \cdot (y - h)]$; $h = 1.3$; $\alpha = 2.754$; $n = 53$

$\beta = \alpha / \log(10) = 1.19 \pm 0.17$. We see that the tail of tornado victims is close to a heavy one (expectation is finite, but variance is infinite). Taking into account that the confidence interval for parameter β

$$(1.19 - 0.17;\ 1.19 + 0.17)$$

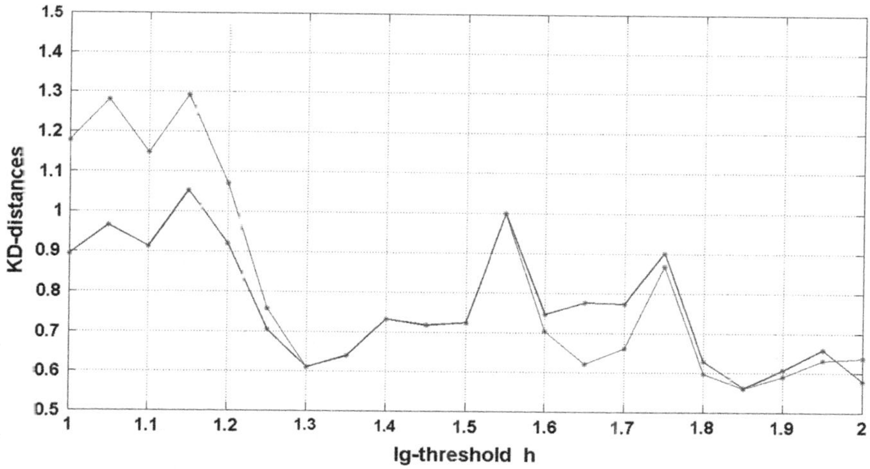

Fig. 3.65 Tornado fatalities, USA, 1953–2012. KD-distances for a grid of log-thresholds. *Thick line*—GPD-fitting; *thin line*—ED-fitting

contains the true parameter value only with probability 84 % (it is valid for the Gauss distribution), one cannot exclude possibility that the true $\beta < 1$, i.e. the tail is really heavy.

Figure 3.66 shows the contribution of the *p*-fraction of the most deadly events to the whole death toll. We see that 10 % of the most deadly events are responsible

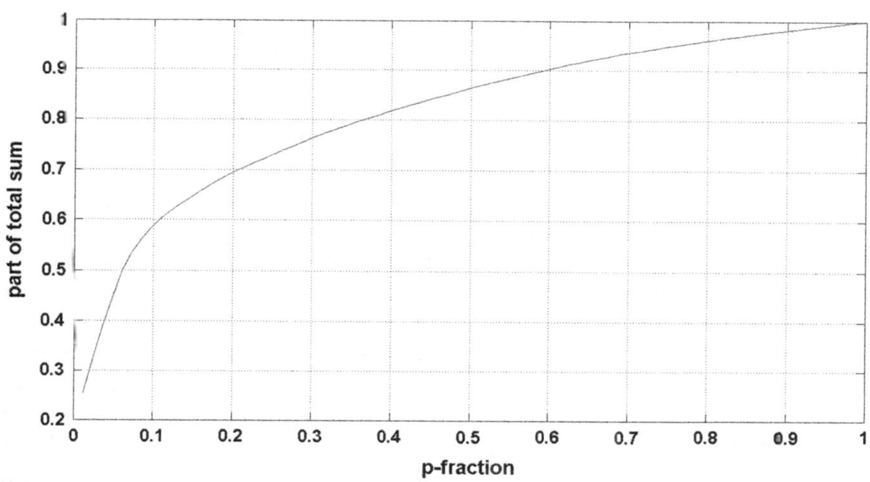

Fig. 3.66 Tornado fatalities, USA, 1953–2012, threshold $h \geq 10$. The contribution of the *p*-fraction of the most deadly events to the total death toll

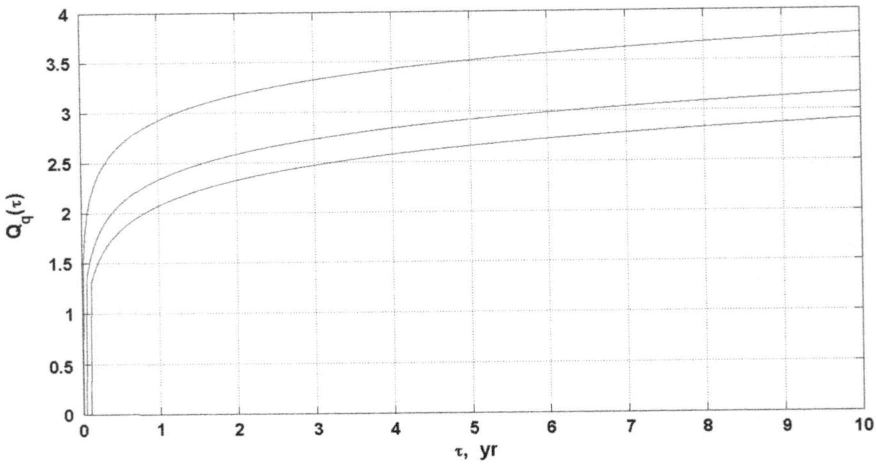

Fig. 3.67 Tornado fatalities, USA, 1953–2012, threshold $h \geq 10$. The ED-quantiles for three different confidence levels: $q = 0.90$ (*lower curve*); 0.95 (*middle curve*); 0.99 (*upper curve*)

for 60 % of the death toll. Such sample can be characterized as sample with a "weak concentration". Figure 3.64 shows the extreme part of the tail used for parameter estimation along with fitted ED-curve. We see that the approximation is more or less satisfactory.

Now we are able to calculate the quantiles $Q_q(\tau)$. Figure 3.67 shows the ED-quantiles for 3 different confidence levels $q = 0.90$; 0.95; 0.99.

3.5 Annual Economic Losses from Floods, USA

In this section we analyze the annual flood damage data compiled by the US National Weather Service (www.flooddamagedata.org). We take the time period 1903–2011. The time series of flood damages in log-scale is shown on Fig. 3.68. The damage is measured in $\$10^9$ (adjusted to 2011). We see that behavior of the time series noticeably changes its character somewhere near 1940. This conclusion is supported by the graph of cumulative sums of log-damage shown on Fig. 3.69. A more or less stable trend is established only after 1940. Thus, we took for further analysis the time period 1940–2011. The sample tail 1 – F(x) of these data is shown on Fig. 3.70. We see rather moderate power-like decreasing (straight line at the extreme range). Figure 3.72 shows the contribution of the p-fraction of the most costly years to the total damage. We see that 10 % of the most costly years are responsible only for 40 % of the total damage. Such sample can be characterized as sample with a "weak concentration". Figure 3.71 shows the extreme part of the tail used for parameter estimation along with fitted GPD-curve. We see that the approximation is quite satisfactory.

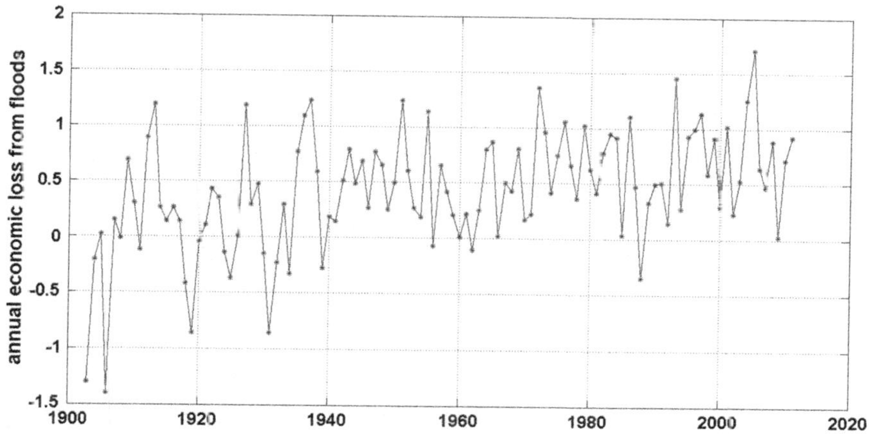

Fig. 3.68 The time series of annual flood damages in log-scale, USA, 1903–2011. The damage is in $\$10^9$ (adjusted to 2011)

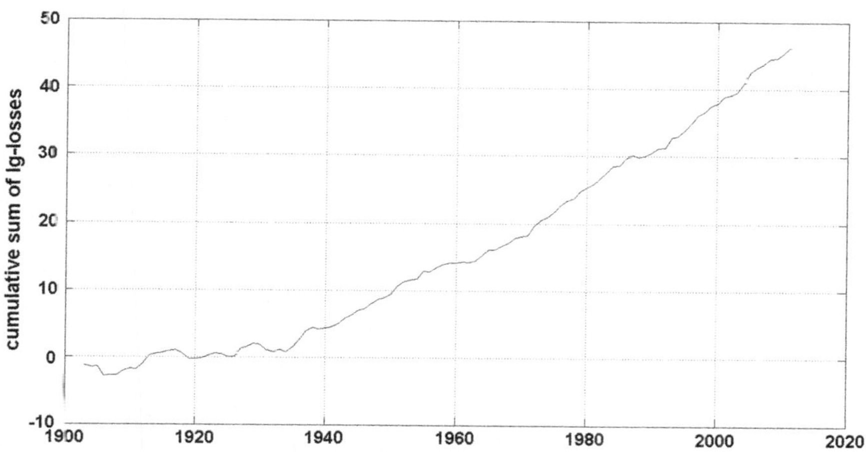

Fig. 3.69 Annual flood damages in the USA 1903–2011. The cumulative sums of log-damages

Now we apply GPD-fitting to peaks over thresholds h. The resulting KD-distances characterizing goodness-of-fit are shown on Fig. 3.73. The best GPD-fitting occurs at $h = 0.4$ ($n = 48$; $\xi = -0.354 \pm 0.093$; $s = 0.541 \pm 0.126$; p–$v = 0.90$). The Q-quantiles in log-scale given by Eq. (2.44) are shown on Fig. 3.74.

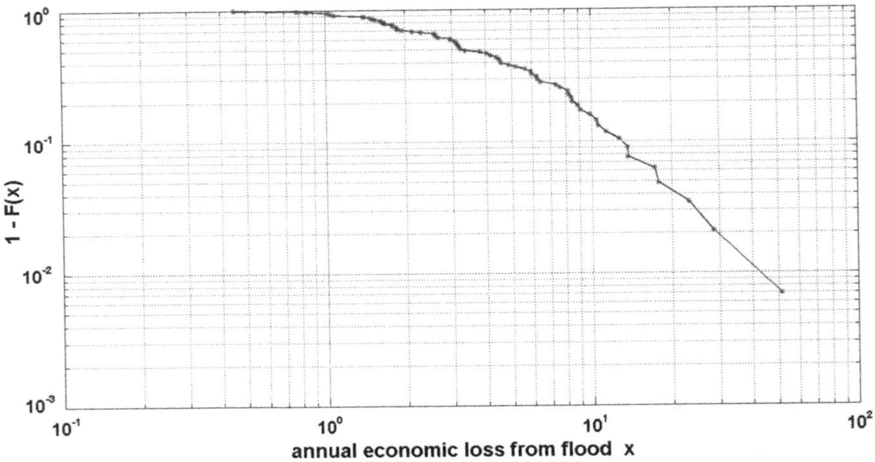

Fig. 3.70 Annual flood damages in the USA 1940–2011. The sample tail $1 - F(x)$

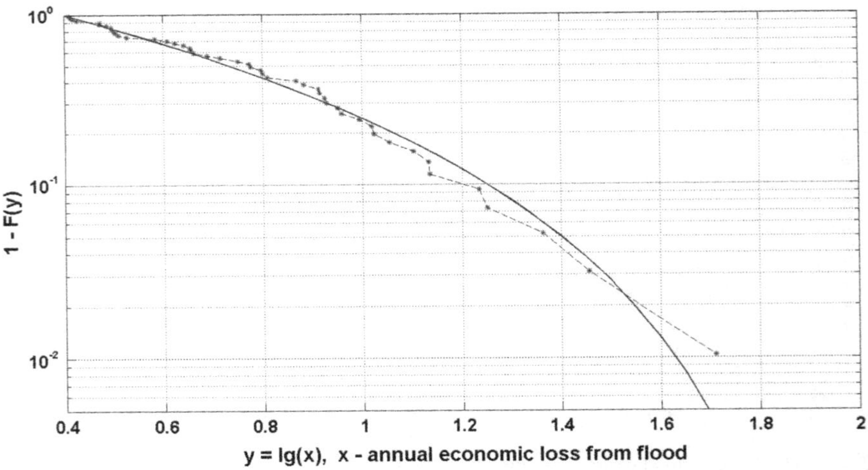

Fig. 3.71 Annual flood damages in the USA 1940–2011. The extreme tail $1 - F(x)$ and approximating GPD-tail: $h = 0.4$; $\xi = -0.354$; $s = 0.541$; $n = 48$

3.6 Annual Economic Losses from Hurricanes, USA

In this section we analyze the annual hurricane damage data published in Blake and Gibney (2011). We take the time period 1940–2010. The time series of hurricane damages in log-scale is shown on Fig. 3.75. The damage is measured in

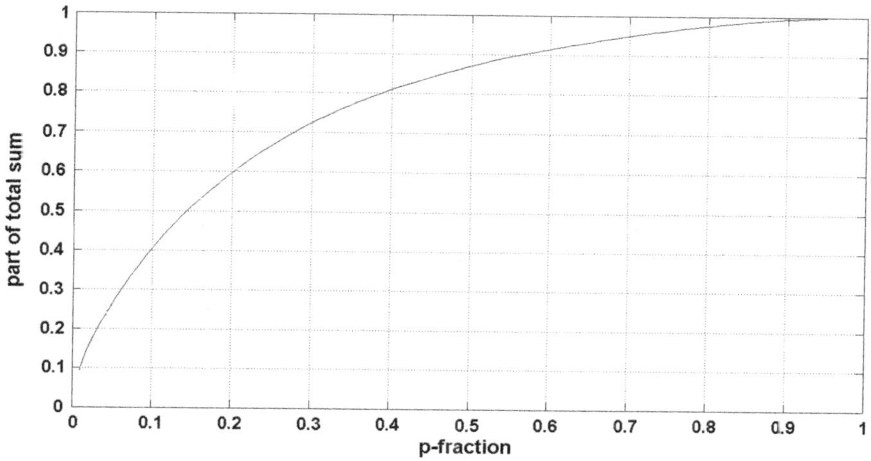

Fig. 3.72 Annual flood damages in the USA 1940–2011. The contribution of the *p*-fraction of the costliest events to the total sum

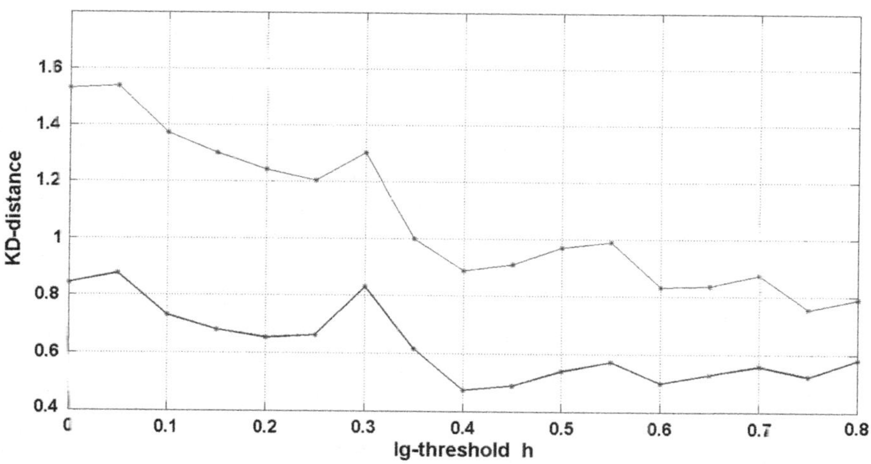

Fig. 3.73 Annual flood damages in the USA 1940–2011. KD-distances for a grid of log-thresholds. *Thick line*—GPD-fitting; *thin line*—ED-fitting

$\$10^6$ (adjusted to 2010). We see that the time series does not exhibit non-stationarity. This conclusion is supported by the graph of cumulative sums of log-damage shown on Fig. 3.76. A more or less stable trend supports assumption of stationarity. The sample tail $1 - F(x)$ of these data is shown on Fig. 3.77. We see irregular decreasing with some fluctuations. Figure 3.79 shows the contribution of the *p*-fraction of the most costly years to the total damage. We see that 10 % of the

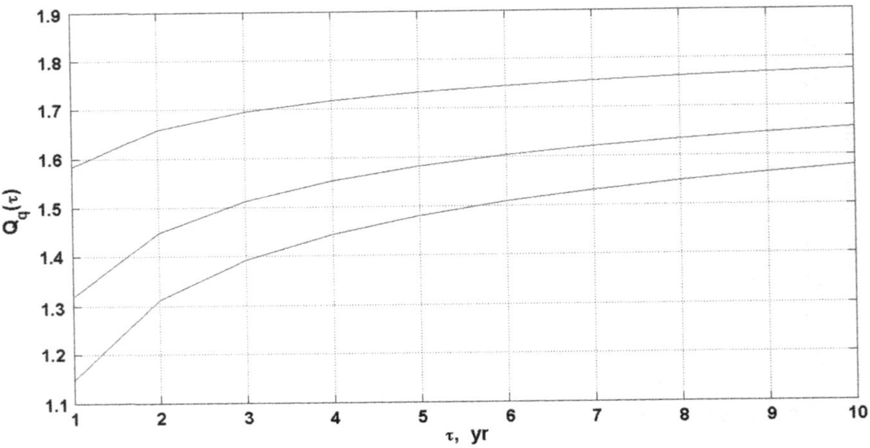

Fig. 3.74 Annual flood damages in the USA 1940–2011. The GPD-quantiles for three different confidence levels: $q = 0.90$ (*lower curve*); 0.95 (*middle curve*); 0.99 (*upper curve*)

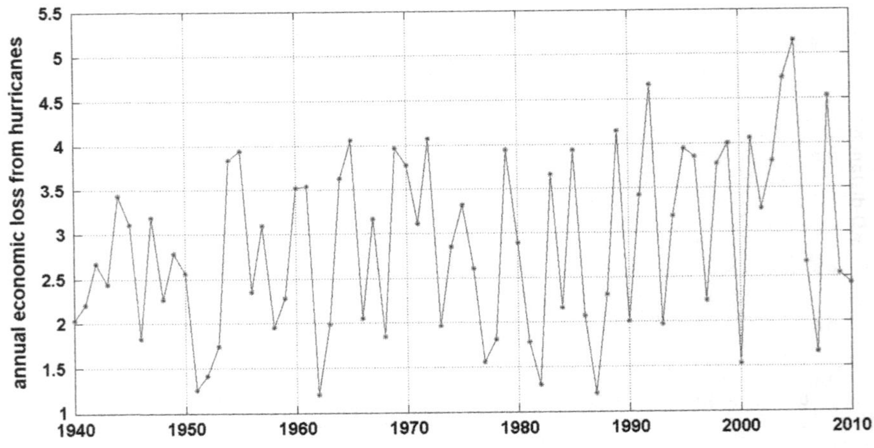

Fig. 3.75 The time series of annual hurricane damages in log-scale, USA, 1940–2010. The damage is in $\$10^6$ (adjusted to 2010)

most costly years are responsible for 70 % of the total damage. Such sample can be characterized as sample with a "moderate concentration". Figure 3.78 shows the extreme part of the tail used for parameter estimation along with fitted GPD-curve. We see that the approximation is satisfactory although there are certain deviations.

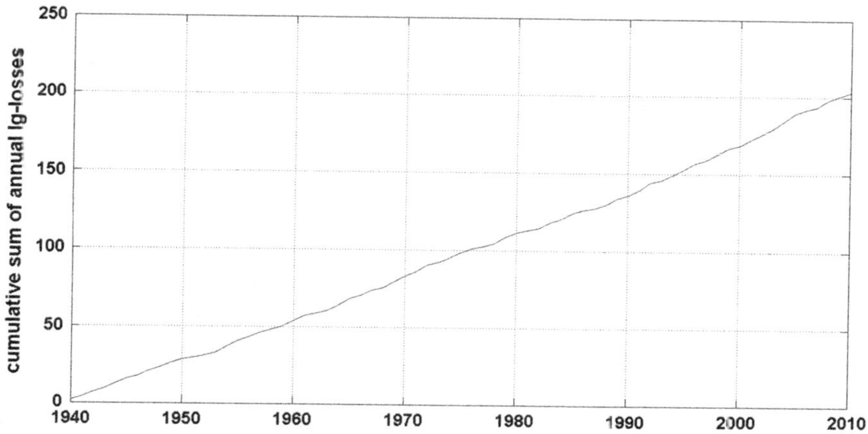

Fig. 3.76 Annual hurricane damages in the USA 1940–2010. The cumulative sums of log-damages

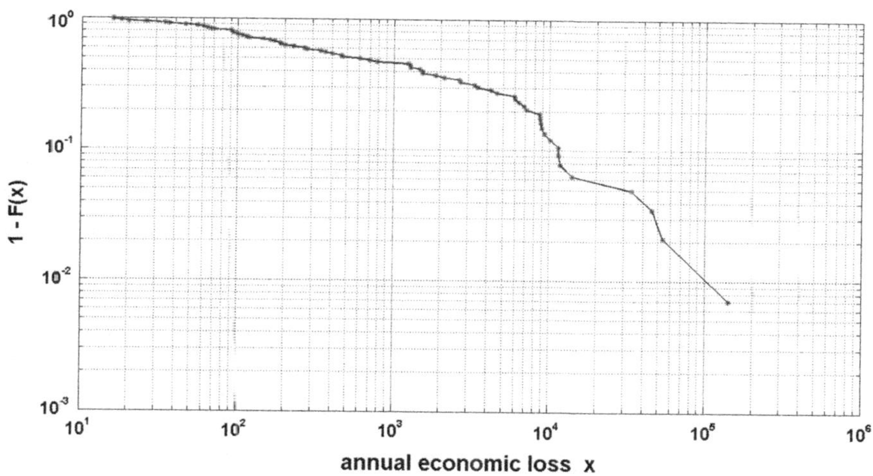

Fig. 3.77 Annual hurricane damages in the USA 1940–2010. The sample tail $1 - F(x)$

Now we apply GPD-fitting to peaks over thresholds h. The resulting KD-distances characterizing goodness-of-fit are shown on Fig. 3.80. The best fitting occurs at $h = 1.5$ ($n = 64$; $\xi = -0.636 \pm 0.045$; $s = 2.368 \pm 0.352$; $p\text{–}v = 0.48$). The Q-quantiles in log-scale given by Eq. (2.44) are shown on Fig. 3.81.

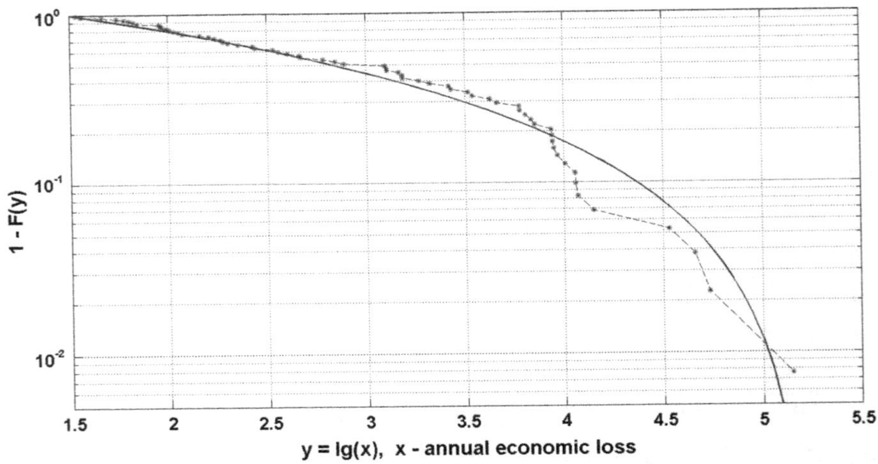

Fig. 3.78 Annual hurricane damages in the USA 1940–2010. The extreme tail $1 - F(y)$ ($y = \log(x)$, x—annual hurricane damage) and approximating GPD-tail: $h = 1.5$; $\xi = -0.636$; $s = 2.368$; $n = 64$

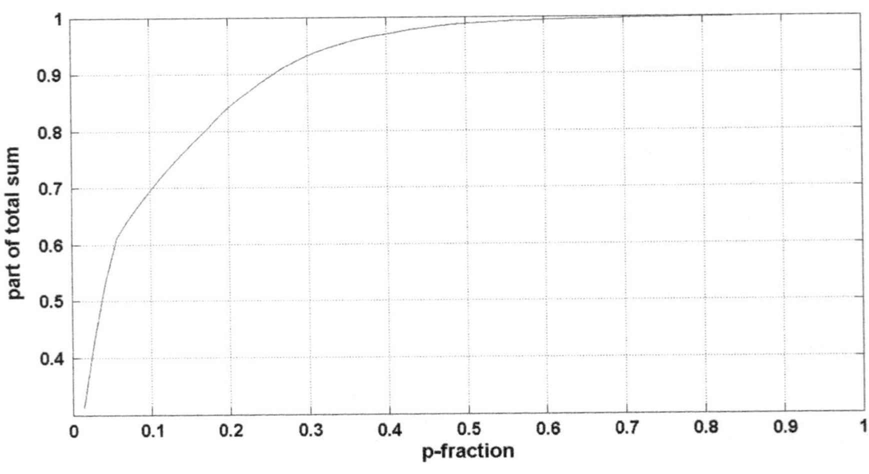

Fig. 3.79 Annual hurricane damages in the USA 1940–2010. The contribution of the p-fraction of the costliest years to the total sum

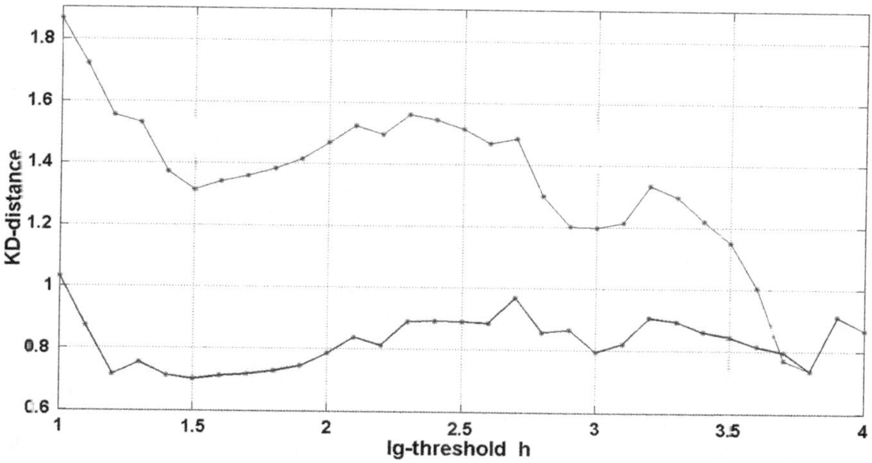

Fig. 3.80 Annual hurricane damages in the USA 1940–2010. KD-distances for a grid of log-thresholds. *Thick line*—GPD-fitting; *thin line*—ED-fitting

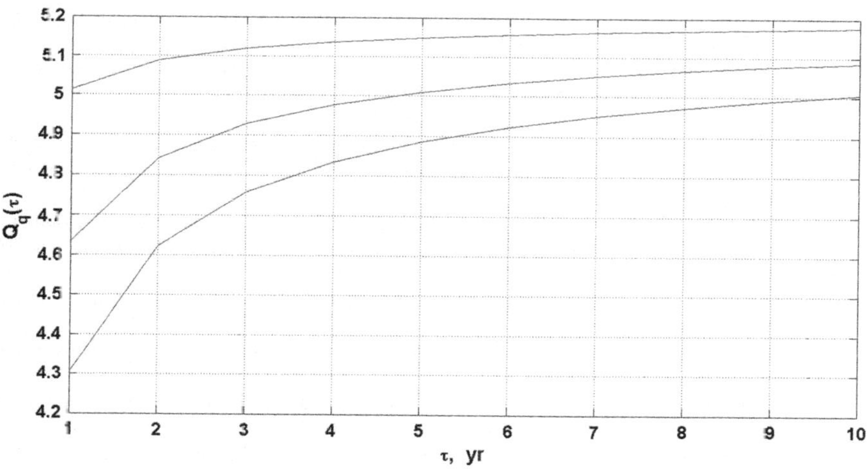

Fig. 3.81 Annual hurricane damages in the USA 1940–2010. The GPD-quantiles for three different confidence levels: $q = 0.90$ (*lower curve*); 0.95 (*middle curve*); 0.99 (*upper curve*)

References

Aptikaev FF (2009) Review of empirical scaling of strong ground motion for seismic hazard analysis. In: Trifunac MD (ed) Selected topics in earthquake engineering-from earthquake source to seismic design and hazard mitigation. Republika Srpska, pp 26–54

Blake B, Gibney E (2011) The deadliest, costliest, and most intense United States tropical cyclones from 1851 to 2010. NOOA Technical Memorandum NWS, NHS-6, August 2011, pp 1–47

Campbell KW (1981) Near-source attenuation of peak horizontal acceleration. Bull Seism Soc Am 71:2039–2070

Cornell CA (1968) Engineering seismic risk analysis. Bull Seism Soc Am 67:1173–1194

Graizer V, Kalkan E (2011) Modular filter-based approach to ground motion attenuation modeling. Seismol Res Lett 82:21–31

Kijko A, Sellevoll MA (1989) Estimation of earthquake hazard parameters from incomplete data files, Part I. Utilization of extreme and complete catalogues with different threshold magnitudes. Bull Seism Soc Am 79:645–654

Kijko A, Sellevoll MA (1992) Estimation of earthquake hazard parameters from incomplete data files, Part II. Incorporation of magnitude heterogeneity. Bull Seism Soc Am 82:120–134

Knopoff L, Kagan Y, Knopoff R (1982) b-values for foreshocks and aftershocks in real and simulated earthquake sequences. Bull Seism Soc Am 72:1663–1675

Lamarre M, Townshend B, Shah HC (1992) Application of the bootstrap method to qualify uncertainty in seismic hazard estimates. Bull Seism Soc Am 82:104–119

Mahdavian A, Aptikaev FF, Erteleva OO (2005) Ground motion parameters in seismically active zones of Iran. Izvestiya Phys Solid Earth 41(2):114–120

Pisarenko VF, Rodkin MV (2010) Heavy-tailed distributions in disaster analysis. Springer, Dordrecht-Heidelberg-London-New York

Pisarenko VF, Rodkin MV (2013) The new quantile approach: application to the seismic risk assessment. In: Rascobic B, Mrdja S (eds) Natural disasters: prevention, risk factors and management. NOVA Publishers, New York, pp 141–174

Pisarenko VF, Sornette D, Rodkin MV (2010) Distribution of maximum earthquake magnitudes in future time intervals: application to the seismicity of Japan (1923–2007). Earth Planets Space, 62:567–578

Steinberg VV, Saks MV, Aptikaev FF et al (1993) Methods of seismic ground motion estimation (handbook). In: Voprosy Inzhenernoi Seismologii, Issue 34. Moskow, Nauka. pp 5–97 (in Russian)

Stephens MA (1974) EDF statistics for goodness of fit and some comparisons. J Am Statist Ass 68(347):730–737

Usami T (1979) Study of historical earthquakes in Japan. Bull Earthquake Res Inst Univ Tokyo 54(3–4):399–439

Usami T (2002) Table of historical damaging earthquakes in Japan. In: Lee WHK, Kanamori H, Jennings PS, Kisslenger C (eds) International handbook of earthquake and engineering seismology (IASPEI), Part A, CD No. 1. Academic Press, San Diego

Utsu T (1979) Seismicity of Japan from 1885 through 1925—a new catalog of earthquakes of *M* 6 felt in Japan and smaller earthquakes which caused damage in Japan. Bull Earthquake Res Inst Univ Tokyo 54(2):253–308

Utsu T (2002) A list of deadly earthquakes in the World: 1500–2000. In: Lee WK, Kanamori H, Jennings PC, Kisslinger C (eds) International handbook of earthquake and engineering seismology, Part A. Academic Press, San Diego, pp 691–717 (The revised and extended version is available at http://iisee.kenken.go.jp/utsu/index_eng.html)

Zlobin TK, Polets A. Yu (2012) Seismotectonics, deep structure and disastrous earthquakes in Kuril-Okhotsk region. LAP LAMBERT Academic Publishing GmbH & Co., KG Saarbrucken, pp 1–93 (In Russian)

Chapter 4
Discussion and Conclusions

We have demonstrated on several data sets related to natural disasters of various nature that using logarithms of the original observations is more appropriate for fitting of heavy tails. By doing so, power-like tails (in particular those obeying the Pareto law with an arbitrary index) are transformed into exponential tails, and the corresponding GPD form parameter becomes non-positive. Zero value of the GPD form parameter corresponds to the exponential tail, whereas its negative values correspond to a distribution with a finite end point M_{max}. Tails heavier than any power-like tail are not frequently encountered in practice, so for the log-transformed data it is sufficient to consider GPDs with non-positive indexes. Thus, the peak-over-threshold distributions of log-sizes of events are best approximated by the GPD with a negative parameter (see Tables 4.1, 4.2). The density function of such distributions takes very small values at the approach of its final point M_{max}, which results in a "duck beak" shape, see Fig. 2.2. For instance, the limit behavior of probability density function of earthquake magnitudes taken from the Harvard catalog is best approximated by the following power law: $(M_{max} - x)^{-1-1/\xi} \cong (M_{max} - x)^{5.14}$. This fact explains in particular the origin of unstable statistical estimates of the parameter M_{max}: smalls changes in earthquake magnitudes can result in significant fluctuations of the corresponding estimates of M_{max}. In contrast, estimates of the integral parameter $Q_\tau(q)$ are typically stable and robust, as we have demonstrated above.

We would like to emphasize that a reliable estimation of quantiles of levels $q > 1 - 1/n$ can be obtained only with some additional assumptions on the behavior of the distribution's tail. Sometimes, such assumptions can be made on the basis of physical processes behind the studied phenomena. Here we have used for this purpose certain theorems of the extreme value theory (EVT). In our case, these EVT based assumptions boil down to assuming a regular behavior of the tail $1 - F(m)$ of the distribution of sizes of events in the vicinity of its rightmost point M_{max}. It should be noted that the assumptions regarding the asymptotic behavior of the distribution's tail cannot equally apply to all practical cases, and they should be supported by additional information for each particular studied phenomenon. In fact, the EVT suggests a statistical methodology for the extrapolation of quantiles

V. F. Pisarenko and M. V. Rodkin, *Statistical Analysis of Natural Disasters and Related Losses*, SpringerBriefs in Earth Sciences,
DOI: 10.1007/978-3-319-01454-8_4, © The Author(s) 2014

beyond the data range; whether such an extrapolation is justified should be thoroughly investigated in each particular case. In our view, the EVT provides us with the best statistical approach to this problem.

Application of the EVT to different extreme events data is reduced to fitting of the GPD to the tail of the corresponding distribution of event sizes or their logarithms. According to the EVT, the Generalized Pareto Distribution is the only possible limit distribution for the "peaks over threshold" events. GPD is a flexible two-parametric family of densities with well-known statistical properties. In certain cases however, even the GPD fails to reasonably approximate the distribution's tail. This may happen in a case when the Limit Theorem of the EVT is inapplicable to a particular data set, since the behavior of the sample's DF in the extreme range cannot be described by a single asymptotic function. For example, it may switch from a power-law like behavior for a certain range of values to an exponential one for the next range of values. In such cases, we have no well defined criteria to choose the value of the threshold for "the peaks over threshold" method, and the application of the exposed approach is not recommended.

Tables 4.1 and 4.2 summarize the main characteristics of the natural disasters analyzed above, together with the parameters of the corresponding fitted GPDs. The first column of Table 4.1 we indicates whether the log-transform was applied to the original values. The third column contains the estimates of the form parameter of the GPD. In two cases the form parameter is null, which corresponds to the exponential distribution (exponential distribution is the limit case of the GPD when $\xi \rightarrow 0$). In all the other presented cases, the form parameter estimates are negative, which indicates the finiteness of the corresponding distributions.

In the fourth column we give the p values which represent the probability to exceed the discrepancy between the observed and the fitted distributions, also known as, the Kolmogorov distance. We consider that if the p value is less than 0.1 one has grounds to reject the fitted curve). One can see that the GPD approximates reasonably well the extreme parts of the distribution's tail for all the considered catalogs of natural disasters. Only in one case (fatalities from floods in USA, 1995–2011) the p value is less than 0.4 which indicates a poor quality of fit. There are two cases (economic losses resulting from floods in USA) when the p value equals 0.90 which corresponds to a very close approximation.

As discussed above, the absolute value $|\xi|$ indicates the steepness of decrease of the extreme part of the distribution's tail. According to Tables 4.1 and 4.2, the steepest extreme tails are observed for the economic losses produced by floods and hurricanes, whereas the corresponding fatality and the injured/affected distributions have, as a rule, smaller parameter $|\xi|$, which corresponds to a slower decay of the tail. As was previously noted, the (unlimited) exponential distribution of $\log(x)$ corresponds to the (unlimited) Pareto distribution of x. This situation occurred once (the last row of Table 4.1) for the case of tornado related fatalities in USA. It is obvious that the maximum number of fatalities in any disaster is limited, however in that particular case a more accurate statistical approximation is observed for an unlimited model.

Table 4.1 Characteristics of disasters and parameters of fitted GPD law

	Lower threshold h, sample size n, intensity λ (1/year)	Form parameter ζ	Goodness-of-fit (p value)	Maximum observed effect	Quantile $Q_{0.95}(10)$
Seismic moment magnitude m_W, Harvard catalog 1976–2012	$m_W \geq 6.8$ $n = 324$ $\lambda = 8.80$	–0.163 ± 0.076	0.59	$m_W = 9.1$	9.13
Earthquake fatalities, Japan, 1900–2011 lg(x)	h = 3 persons $n = 44$ $\lambda = 0.339$	–0.260 ± 0.111	0.43	142,807 persons	58,000 persons
Injured in earthquakes, Japan, 1900–2011 lg(x)	h = 3 persons $n = 99$ $\lambda = 0.884$	–0.374 ± 0.063	0.69	103,733 persons	75,000 persons
USA, perished in floods, 1995–2011 x	h = 3 persons $n = 41$ $\lambda = 1.11$	0.0	0.22	35 persons	53 persons
Affected in floods, USA, 1995–2011 lg(x)	h = 500 persons $n = 52$ $\lambda = 3.06$	–0.182 ± 0.113	0.66	11,000,148 persons	17,700,000 persons
USA, estimated economic losses from floods, 1995–2011 in millions of $, lg($x$)	$h = 80$ $n = 32$ $\lambda = 1.88$	–0.486 ± 0.091	0.90	12,000	12,400
USA, perished in tornadoes, 1953–2012 lg(x)	h – 20 persons $n = 53$ $\lambda = 0.88$	0.0	0.75	1,200 persons	1,480 persons

Table 4.2 Characteristics of annual disasters and form parameter of fitted GPD-law

	Lower log-threshold h (10^h) sample size n	Form parameter ξ	Goodness-of-fit (p-value)	Maximum observed effect, lg(x) (x)	Quantile $Q_{0.95}(10)$ (10^Q)
Annual economic losses from floods in USA, in 10^9 \$, 1940–2011 $\lg(x)$	h = 0.4 (2.5) n = 48	−0.354 ± 0.093	0.90	$\lg(x) = 1.71$ (51.3)	1.66 (45.7)
Annual economic losses from hurricanes in USA, 1940–2010 in 10^6 \$, $\lg(x)$	h = 1.5 (31.6) n = 64	−0.636 ± 0.045	0.48	$\lg(x) = 5.15$ (141,000)	5.09 (123,000)

One can observe that in certain cases the quantile $Q_{0.95}(10)$ is less than the observed maximum event size, while in certain other cases it exceeds that value.. This is a result of an interplay between the parameters of the fitted GPD, namely intensity λ and time interval τ. It should also be remarked that such characteristics as economic losses resulting from natural disasters are strongly influenced by a rapid global development of the economic infrastructure and the population growth. Therefore, it is quite difficult to reliably forecast such characteristics for long time spans, say beyond 10–15 years. This remark should be kept in mind when one estimates quantiles of future losses.

Table 4.2 summarizes the results of the analysis of annualized data. The aggregation of event sizes over one year intervals represents in essence a linear filtration (smoothing) of the corresponding time series of sizes. That is why the tails of annualized distributions are as a rule less heavy compared to the tails of original distributions of marked point processes. This fact can explain higher values of the form parameter (in terms of its absolute value) of annualized distributions in Table 4.2 compared to the corresponding form parameters in Table 4.1. One exception is the case of the economic losses from floods, which can be explained by a very small sample size in this case: $n = 32$ (single event losses) and $n = 48$ (annualized losses). We remind that the theoretical maximum M_{max} of the GPD distribution with negative form parameter ξ is expressed as

$$M_{max} = h - \frac{s}{\xi},$$

and the lesser $|\xi|$ the larger M_{max} is.

One can also note, that the correlation between the high quantile $Q_{0.95}(10)$ and the maximum observed size is stronger for the annualized data, as it could be expected.

Table 4.3 Ratio of sum of 10 % largest effects to total sum

	Ratio of sum of 10 % largest effects to total sum (%)
Affected in floods, USA, 1995–2011	98
Earthquake fatalities, Japan, 1900–2011	98
Injured in earthquakes, Japan, 1900–2011	94
Annual economic losses from hurricanes in USA, 1940–2010	70
USA, estimated economic losses from floods, 1995–2011	68
USA, perished in tornadoes, 1953–2012	60
USA, perished in floods, 1995–2011	55
Annual economic losses from floods in USA, 1940–2011	40

We gave in Chap. 1 theoretical relations (1.3)–(1.4) connecting the sample maximum $M_{\max}^{(n)} = max(x_1, \ldots, x_n)$ with the total sum $S_n = x_1 + \cdots + x_n$. We can as well compare S_n with the sum of k largest observations. The ratio of such sums for the analyzed catalogs is presented on Figs. 3.27, 3.35, 3.42, 3.50, 3.59, 3.64, and 3.70. These ratios reflect in a more in detailed manner the contributions of the rightmost part of tail to the total sum. Let us consider for comparison one particular value on these curves, namely the ratio of 10 % of the largest observations to the total sum. One can say, that the higher this ratio, the more events are concentrated around the tail's extreme range. Table 4.3 presents a collection of such ratios for all the considered event catalogs. One can conclude that the highest concentration of events around the distribution's tail is observed for the data sets related to the number of individuals affected by floods (USA), to earthquake fatalities (Japan) and to the injured by earthquakes (Japan). For these cases, 10 % of the largest events are responsible for more than 95 % of the total loss. Intermediate values of the event concentration toward the tail's end (about 60–70 %) are observed for annualized economic losses from hurricanes (USA), economic losses from floods (USA) and fatalities from tornadoes (USA). Weak concentration (40–55 %) is observed for flood fatalities (USA) and annualized economic losses from floods (USA). It should be noted, that our concentration graphs are in essence an extended analog of the Pareto principle (or the 80-20 rule): "for many phenomena roughly 80 % of the effects come from 20 % of causes" (Italian economist Vilfredo Pareto observed in 1906 that 80 % of the land in Italy was owned by 20 % of population).

The manufacturer's authorised representative in the EU is Springer Nature Customer Service Centre GmbH, Europaplatz 3, 69115 Heidelberg, Germany. If you have any concerns regarding our products, please contact ProductSafety@springernature.com

Printed and bound by CPI Group (UK) Ltd, Croydon, CR0 4YY

05/10/2026

02228151-0002